THE ASCENT OF HUMANITY

Also by Amit Goswami

The Self-Aware Universe

Science and Spirituality

The Visionary Window

Physics of the Soul

The Quantum Doctor

God Is Not Dead

Creative Evolution

How Quantum Activism Can Save Civilization

Quantum Creativity

Quantum Economics

The Everything Answer Book

Quantum Politics

See the World as a Five Layered Cake

With Valentina R. Onisor

Quantum Spirituality

The Quantum Brain

The Quantum Reenchantment of the Reality You Live

Hero's Journey: Quantum Style

Quantum Integrative Medicine

The Return of the Archetypes

With Sunita Pattani

Quantum Psychology and Science of Happiness

With Carl D. Blake and Gary Stewart

Quantum Activation: Changing Obstacles into Opportunities

The Ascent of Humanity

QUANTUM BIOLOGY AND OUR FUTURE

Amit Goswami, PhD

LUMINARE PRESS

WWW.LUMINAREPRESS.COM

The Ascent of Humanity: Quantum Biology and Our Future
Copyright © 2024 by Amit Goswami, PhD

Although every precaution has been taken to verify the accuracy of the information contained herein, the author and publisher assume no responsibility for any errors or omissions. No liability is assumed for damages that may result from the use of information contains within.

Cover Design by Debabrata Dey Biswas
Printed in the United States of America

Luminare Press
442 Charnelton St.
Eugene, OR 97401
www.luminarepress.com

LCCN: 2024926168
ISBN: 979-8-88679-743-5

To all Quantum Aficionados of the world.

Together, let's make it a better place

Table of Contents

ACKNOWLEDGEMENT

I would like to thank Valentina Onisor and Krishanu Goswami for careful reading of the manuscript. Thanks are also due to the Masters and PhD students at our Quantum Activism Vishwalayam and University of Technology, Jaipur, for their enthusiastic reception of the message of this book.

PROLOGUE

Quantum Biology, Creative Evolution, and the Future of Humanity

Quantum physics gives a new metaphysics—consciousness is the ground of all being—for science. Accordingly, in my previous books, I have already provided the basics of a new paradigm of quantum science and applied it to psychology and medicine. I did make an early attempt to formulate a quantum biology; however, it did not fully solve and integrate all the new questions and ideas that maverick biologists have suggested from time to time toward the solution some of the important unsolved problems of biology:

- The nature of life—how life is differentiated from nonlife.

- The origin of life given the circularity—it takes DNA to make the protein but it also takes protein to make DNA—in life's production.

- Two tempos of evolution: there is continuous evolution of slow tempo; there is also rapid evolution of fast tempo; the fossil records are continuous except for conspicuous gaps.

- Cell differentiation in organs: all cells of an organism carry the same genes, but each organ activates only the genes that are required for that organ's functioning.

- How purposive development contributes to evolution as indicated by the progression from simple organisms to complex orgainisms signifying a biological arrow of time.

- How evolution is really evo-devo—evolution and development occurring together.

- Directed mutation: how lowly bacteria, under threat of starvation, directs its own gene mutation rate to survive.

- The general question of animal consciousness—how can animals be conscious without a neocortical brain?

- How humanity broke free from survival-directed evolution and turned its attention to higher needs.

In this book, I have developed a complete solution of all these unsolved problems with a quantum biology based on quantum physics and the primacy of consciousness.

The ongoing theme of the book is the idea of the ascent of humanity, how humanity broke free from the shackle of survival motif alone and turned its attention to embodying the archetypal values of consciousness. This theme is usually avoided by professional biologiests but is of utmost importance for readers who believe in the human potential movement. Accordingly, I have written the book with the nonspecialist reader-friendliness in mind.

We are in the twenty-first century and humanity is faced with repeated crisis involving global climate change, terrorism including state terrorism, economic meltdowns, the breakdown of democracy and the rise of totalitarianism, etc. You can sum it all up by saying: we are seeing a massive decline of human civilization!

There are many doomsday scenarios and conspiracy theories going around. And yet, not so long ago, the philosophers Sri Aurobindo (1996) and Teilhard de Chardin (1961) theorized that evolution is guided by conscious purpose and predicted a brilliant future for the humankind: *heaven on earth.* In view of the current crisis, can one still be optimistic?

What lies in our future? I believe the current crisis is nothing but a wake-up call. Its purpose is to wake us up from what the poet William Blake called "Newton's sleep", wake us up to meaning and purpose.

Let me elaborate. Most scientists today are glued to Darwin's theory of evolution (Darwin, 1857) in the revised form of Neo-Darwinism: in this view, we are animals driven by the survival motif; the future of evolution can go either way, decline or ascent depending on how we see it, but we will still be animals trying to survive. Yet, there are much fossil data indicating that a re-vision of Darwin's theory big time is necessary, although it is not politically correct for a biologist to admit it. That would only add fuel to Christianity's tendency of negating evolution altogether, the biology establishment defensively points out.

To avoid religions and their dogmas, most scientists today insist that science must be done wearing a "straitjacket," of scientific materialism—the ideas that 1) all is matter, 2) matter at the macrolevel where we live follow Newtonian deterministic laws, and 3) we are biorobots with ornamental experiences that we call consciousness. Unfortunately, thinking we are biorobots we tend to become that. In this way, this worldview can be dangerously self-fulfilling.

That label 'straitjacket' was given by one of the truly great scientists of the last century—Richard Feynman. Is the idea "everything is matter" really a straitjacket for science to wear? Indeed so, it is a dogma; dogmas have no place in science whose avowed purpose is to find the truth about reality. Go figure.

When biologists wore this straight jacket, biology embraced molecular biology—the idea that biology is chemistry. There is no difference between the nonliving and the living; the apparent sentience of living beings can be ignored as causally ineffective.

In the true spirit of science, we should consider scientific materialism and molecular biology as tentative hypothesis to be tested by experimental data. Experimental data has spoken, and the verdict is negative (see later). Unfortunately, the stranglehold of scientific materialism and molecular biology over practicing scientists continues.

Humanity is caught between two battling dogmas—religion and science. The religious idea that God created the world and all life in it in one fell swoop is as dogmatic as scientific materialism's claim that matter creates life and consciousness. Both dogmas take away our freedom to explore meaning and purpose and the prerogative to act appropriately. Religions are stuck with the idea that heaven comes to us only after death, it is not achievable here on earth; scientists preach humans are robots with ornamental experience; they have no freedom, no values to transform with; they are stuck with their pleasure-seeking, negative emotions, and me-centeredness of their animal ancestry.

To solve crisis, we need creative people to come together and explore new ideas. The polarization of worldviews has made this difficult. And yet, everyone should know, materialist science with dogma does not reflect the true spirit of science. And religions not only deny scientific data on evolution but also render the undeniable human experiences of spirituality into the pursuit of unnecessary dogmas in the service of their inadequate explanatory principles. *People should recognize that both sides are engaged in a power play, not search for truth.*

There is fortunately, light at the end of the tunnel of competing dogmas. Since the early nineteen-hundreds, a paradigm shift has been taking place in science. With the formulation of quantum science within the primacy of consciousness (Goswami, 1993, 1994) and further research applying this new paradigm in biology, psychology, and health sciences is slowly but surely replacing Newtonian/Darwinian science.

The most striking difference between Newtonian physics and quantum physics is this: in Newtonian physics objects move in one and only domain of space and time. In quantum physics, objects are possibilities of consciousness to choose from in a domain of reality outside space and time—the domain of potentiality; only with observation, the possibilities collapse to actual events of the observer's experience in the domain of space and time.

Quantum physics is dogma free. Contrary to what you read in popular science media, it is also paradox-free. Only if you look at it with the materialist dogma, you get paradoxes.

Quantum science based on quantum physics is also dogma free as well as paradox-free; its metaphysics—that the domain of potentiality that quantum physics defines is the nondual oneness of consciousness defined by spiritual wisdom traditions—is experimental metaphysics. It has been verified experimentally already (Grinberg et al, 1994, Standish et al, 2004).

Quantum science brings consciousness and conscious purpose into science. It gives us an unexpected avenue to explore the most fundamental human experience—conscious awareness with its two poles—the experiencer and the object of experience, the subject "I" and the object "it". Importantly, it also integrates science and spirituality. This in its turn makes it possible to integrate spiritual theories of evolution (Aurobindo, de Chardin) with scientific theories (Lamarck, Darwin).

Accepting the dogma of molecular biology, biologists have missed the biggest story of evolution—the progressive evolution of the embodiment of consciousness and its ideas. Life begins when consciousness is embodied in nonliving matter in the simplest possible way in the form of a single living cell—consciousness manifest at the microlevel of being. From single cell creatures came multicell creatures came fungi, plants and a variety of animals with locomotive capacity. Still, consciousness remained in the microlevel, difficult to recognize.

What was not difficult to recognize is that evolution of the multicellular is driven by the survival motif as Darwin theorized. Astute philosophers recognize that survival is an idea of consciousness and not a property of molecules, however complex. But molecular biologists theorized that the macromolecules of the living—DNA, RNA, and protein—*somehow* develop survivability and therefore, molecular biology is synchronized with Darwin's theory of evolution. In this way, biologists developed the belief that Darwin's theory confirms scientific materialism.

About a hundred million years ago, evolved the mammals with a brain large and complex enough to enable cognition; they seem to have consciousness at the macrolevel, at their brain. Still, the potentialities opened up via this ability was not immediately clearcut. Only much later, with the development of humans, did the nature of these potentialities became clear as they manifested in a new phenomenon of experience—domestication. Human were able to first domesticate themselves (self-domestication), and second domesticate a whole bunch of other mammals, even birds, although not that big a bunch (Wrangham, 2019).

Domesticity is not necessary for survival; the wild cousins of domestic animals survive quite well. The survival motif is not enough to explain living phenomena when self-domestication and later domestication came into the scene.

Self-domestication took humans beyond the explanatory power of today's establishment biology. Similarly, domesticated animals are also beyond normal survival biology. Domesticated creatures dance to higher needs in addition to their survival needs; their behavior is driven by not only cause but also by purpose. A new quantum biology of purpose is needed.

Spiritual traditions have been telling us about purpose of human life, higher needs, and human spiritual values for millennia. In modern psychology, Abraham Maslow integrated the idea of survival needs with higher needs by defining the hierarchy of needs—when survival needs are satisfied, one looks for the satisfaction of higher needs.

Self-domestication enabled humans to venture into the exploration of higher needs; humans did it for themselves and by domesticating a few other mammals demonstrated that some of the higher mammals—dogs, cats, horses, etc.— have this capacity as well.

Biological establishment has created a lot of fog around the idea of survival being a property of the macromolecules of life, but nobody can doubt that higher needs are different from survival needs and entities that strive for higher needs cannot be entities made of matter alone. Since we cannot have two different paradigms of biology, one for creatures of survival need and one for creatures of higher need, we must acknowledge that old establishment biology must be replaced by quantum biology.

Biological theory of evolution has another debate which has never really been settled: Is evolution entirely the product of what takes place at the genetic micro (genotype) level or are developmental changes taking place at the macro (phenotype) level important as well for evolution? What is now called the Lamarckian theory of evolution was proposed by the biologist Jean-Baptiste Lamarck even before Darwin; it is exclusively based on the idea of acquired characteristics during an organism's development. For example, the long necks of giraffe that we see today, Lamarckians would say, are due to generations of giraffes stretching their necks to reach high branches of tree-leaves. But Lamarckian theory could not be taken seriously because nobody could solve the problem of hereditary transfer of somatic developments. Macroscopic developments are driven by purpose; in this way the debate is also a battle between cause and purpose.

I declare: the problem of heredity of macroevolutionary developmental traits has been solved. Emboldened by this, in this book, I will develop a new quantum biology that incorporates both cause and purpose. I will then develop a theory of creative and purposive evolution incorporating both Darwinian and Lamarckian ideas -call it quantum evo-devo (see also, Carroll, 2005).

You will see that the worldview of quantum science when applied to evolution, quantum evo-devo incorporates the ideas of Aurobindo and Teilhard integrating them with Darwin's and Lamarck's ideas as well.

In 2008, I wrote a book *Creative Evolution*, demonstrating that quantum science can explain fossil data better than Neo-Darwinism and yet this science within consciousness incorporates spirituality and therefore, even open-minded religionists should not deny it outright. However, the book was not well received in either establishment, religion or science.

In the meantime, the crisis of civilization has deepened and right now is threatening our very civilized survival. To bring together the two fighting factions—dogmatic science and dogmatic religion—is an utmost priority; we must develop strategies for solving the crisis conditions and we can do that easily if the two factions change their dogmatic positions and accept this worldview integrating new paradigm biology and evolution theory.

I show in this book that the theory of quantum evo-devo that incorporates purpose from the get-go shows us the way out from the current crisis situation. The theory shows you how even a relatively small number of dedicated people working with quantum principles of transformation can bring about the necessary change to take us beyond the crisis. The recipe is really quite simple in principle: if this small number of dedicated people engage creative transformation together using quantum principles in synchrony with the evolution-ary movement of consciousness, the entire humanity, fractionated or not, will effortlessly transform as well. I will demonstrate with evolutionary data that it has happened before, and that we can make it happen again.

Why is this important? The 'we' that will make this momentous change happen are we the authors and you the readers interested in spirituality and transformation. Getting started with transfor-mation requires conviction that comes from clarity—clarity about who are are, where we came from, where we are going.

The Ascent of Humanity: An Intro to the problems and a Glimpse at the solutions

The Case for Quantum Biology—a Biology of Meaning and Purpose—and the Ascent of Humanity

There is now a new consensus developing: human beings, homo sapiens took off and became the only surviving human race because some 70000 years ago, they took the decisive revolutionary step of leaving the exclusive pursuit of the survival need behind and opted additionally for higher needs of meaning and purpose. According to the philosophy of humanism, the purpose—to explore and embody human values such as love and goodness was invented by humanity itself to develop society and civilization. Hence ever since the revolution circa 70000 BCE, human society and civilization is ascending.

Humanism is philosophy, most scientists do not take it seriously. According to establishment biology today, biological evolution is driven by the survival motif, period; humans are no exception. There is no room for purpose. In this view, all the human development that some of us may see as a result of the pursuit of meaning and purpose are nothing but the epiphenomena of the survival motif.

Most professional biologists subscribe to the one hundred and fifty-year old theory of evolution of Darwinism (its revised version of Neo-Darwinism actually; however, the revisions are mostly of technical details). In this theory, there is really no way to gauge ascent or descent. So according to Darwinism, in the future the human civilization on earth could very easily give way to a *planet of the apes* as the classic movie by that name portrays.

This establishment view does not make humanism wrong. Also, that something significant happened circa 70000 BC is based on data, albeit scanty. The truth is, there is plenty of other indications that we need a new biology and an extension of Darwinism to include meaning and purpose. There is another theory of biological evolution by Jean-Baptiste Lamarck based on the propagation of acquired characteristics that has purpose built into it. This theory has data behind it as well, some of it recent (Cairns et al, 1988). In this book we will develop a new quantum biology of meaning and purpose; its centerpiece is a new theory of simultaneous evolution and development—let's call it quantum evo-devo—based on quantum physics and the primacy of consciousness.

Darwin and his subsequent followers all make one fundamental assumption that is unfounded: genes determine form, or in sophisticated words, genotype determines phenotype. Well not entirely. A recent discovery has settled the issue: all cells in a complex animal body have the same genes; but the mechanism that determines which genes are activated in which organ to endow it with form and function is epigenetic—outside of the genes.

Also, undoubtedly, natural selections take place only at the macro phenotype level. This is why in evolution both gene mutation and development of form and function which requires a different mechanism are important; evolution is evo-devo—both Darwinian and Lamarckian mechanisms contribute. Quantum evo-devo accomplishes the integration of both sets of mechanism for evolution.

In contrast to Darwinian mechanisms of chancy gene mutation and natural selection both of which have no directionality in time, in quantum evo-devo, later forms build on earlier form; form is accumulative and progressive in time. In this way, we get the observed biological arrow of time.

Let's consider the idea of ascent of humanity. Sociobiologists following Darwin professes no ascent. Humanists give up on Darwinism for human development and professes that the pursuit of human values leads to the ascent of humanity. Unfortunately, currently, human values are under attack. Most scientists side with the philosophy of scientific materialism—everything is matter and material interactions. In that theory human beings are not capable of embodying values; the best they can is do is pretend to follow values.

According to the anthropologist Richard Wrangham, the revolution that happened in the way homo sapiens lived 70000 years ago was the culmination of the process of self-domestication (see prolog). However, self-domestication, he theorized, gets rid only of reactive aggression not of what he calls proactive aggression via which humans can plan long term wars. In this view, although with self-domestication, we have learned to make bigger societies, we will always be limited because of proactive aggression and make wars between societies.

The scenarios of war-making are getting worse. In the latest version of scientific materialism, the fundamental commodity of the material world is information. In this theory human beings just like silicon beings of artificial robotic intelligence are nothing but information-exchanging bio robots; their experiences—including values mean nothing. And then what may lie in our future is a battle for supremacy between bio-robots—humanity—and silicon robots of AI. Who will win? No can tell (Harari, 2015). Makes you anxious, doesn't it?

Evolution according to quantum biology incorporates Darwinian causal mechanism as well as Lamarckian drive of purpose and in this way gives us a different theory of human values—that human discovery of values has a spiritual origin; values come from aspects of consciousness—the ground of being—called archetypes.

In this book using quantum biology, I will show that the ascent of humanity is about increasing the scope of conscious expression, it is about humanity's ability to explore, embody, and live by the archetypal values.

The revolutionary process of self-domestication that made us humans allowed us to make bigger societies than our ancestors—for example, chimpanzees—could. In quantum science, we see self-domestication as the beginning of the process of embodiment of the archetype of goodness. As we embody goodness more, we develop first altruism and then compassion for the fellow human being. As we live by the archetypes, we develop societies of greater and greater scope.

However, not all people can transform. This sets up a good-evil dynamics in our society that opportunistic leaders manipulate and cause wars. Because of value denigration and polarized worldviews, we are still struggling with the mission of building and holding together nation/states—societies consisting of people of common agreed-upon values. We will have to complete that mission. Then only, we will be able to develop a planetary bond of goodness among the current nation/states. And get rid of wars forever.

Meaning and Purpose

There is story about a philosophy professor who was about to retire. For his final lecture to his class, he decided to talk on the meaning of life. When the students came into the classroom, they were surprised to find the professor sitting at his desk with the desktop filled with stones of large and small sizes as well as sand along with three jars. The professor started to fill one jar with the large stones. When it was full, he asked, "Is the jar full?" The students were puzzled by the antique of the professor but of course they responded yes.

The professor then filled a second jar with the smaller stones and asked the same question that received the same answer. Ditto when he filled the last jar with sand.

Now the professor explained to his class filled with curiosity. "See, the cobbles are about the big things of life like family, jobs, etc. This is where you find your most important meanings. The smaller stones represent mid-level meaningful things of life like job interviews, exams, dating, etc. And the sand is about small things like brushing your teeth, checking your cell phone, etc.

"The way to live a meaningful life is to pay maximum attention to objects of the first category and the least attention to the objects of the last category. If you fill up your life with the last category, then you will be in trouble."

The professor and the students discuss this idea for a while. At the end, the professor brings out a bottle of beer and pours it in the jar containing cobbles. Then he says, "See, however much you fill up your life with things, there is always room for a beer. Remember that."

The professor's advice is useful no doubt, but why the beer? When we categorize meaning processing this way, life soon becomes work and does not bring satisfaction anymore. So, we need the beer to relax from the stress of living. Is there a better way?

There is. We can look at meaning processing somewhat differently. We have four different kinds of experience—sensing, feeling, thinking, and intuiting—and our mind gives meaning to each of these experiences. Indeed, Sri Aurobindo used this idea to categorize human development progressively, and the anthropological data so far is in agreement. In the hunters-gatherers era, people's mind was busy giving meaning to sensory objects of what to hunt and what to gather and hence they developed the physical mind. In the next era of garden agriculture, people lived in families and cultivated feeling and developed the vital mind. Next came heavy agriculture, some people had spare time and they began processing of meaning of meaning itself—abstract thinking. This gradually gave us the era of the rational mind. The next era of the intuitive mind—people mainly giving meaning to intuition—is yet to come.

Even today, we can categorize people into people of physical mind, of vital or emotional mind, of mental or intellectual mind, and of intuitive mind. But here is again, something important to notice. At the end of the day, people of the first three kinds will still need the beer for relaxing.

But something is different for people of the intuitive mind. Intuition brings us the archetypes. People of intuitive minds explore and embody the archetypes and both the means and the end are full of joy and satisfaction. What's the difference? These people have combined their search for meaning with higher purpose, with the purposive movement of consciousness.

In summary, the optimum way to live our meaning life is to explore meaning with higher purpose at least some of the time.

What is Happening to Our Search for Meaning and Purpose Today?

A substantial number of people today are looking for meaning and purpose of their lives. What is downright depressing to them is that the current establishment science burdened with the dogma of scientific materialism—everything is matter—has no room for meaning and purpose.

Although you will never get the idea from the mainstream media, it is no secret that the dogma of scientific materialism with causal determinism to guide it works well only for *macroscopic* nonliving systems of physics and chemistry.

Living systems have new *nonmaterial* potentialities to explore, manifest, and embody. This is what evolution is about. When it comes to humanity, we evolve mostly via development—by developing the capacity of exploring and embodying nonmaterial potentialities of meaning and purpose and transform.

We cannot deny our biological heritage of animal ancestry. We are all born with a survival-oriented base-level human condition defined by a survival-oriented me-centeredness, negative emotional brain circuits., and a tendency for addiction to pleasure.

The biggest damage scientific materialism applied to biology and psychology has done in the last six decades is to create a societal mind-set that accepts the base-level condition as humanity's definition. This has shut off the exploration of meaning and purpose in people's lives.

Starting with the book *The Self-aware universe* (Goswami, 1993), I have been working on a new paradigm of science based on quantum physics and the primacy of consciousness, call it quantum science of consciousness, or simply quantum science. This science of human psyche brings back meaning and purpose and shows their relevance for human physiological and psychological development (Goswami and Onisor, 2024).

In this book, I will show that purpose goes even deeper, to the heart of biology—science of life and evolution—itself. I develop here the quantum biology of meaning and purpose. I show how the purpose is embodied in matter to produce life and how evolution is an evolution of embodied purpose.

As life evolved, initially, the purpose was survival and that could easily seem like a cause leading to a causal interpretation of Darwin's original idea. But as life continued to evolve, we see evidence of exploration of higher purpose all through the animal kingdom. Co-evolution for mutual benefit is an example.

However, humanity in the form of Homo Sapiens was the first to *embody* a higher archetypal purpose in the process of self-domestication defying the biology of survival. Eventually, out of all the human spwcies of the genus Homo, only sapiens were able to survive. *In this way, today, the capacity for the embodiment of higher archetypal purpose defines humanness.*

According to some bio-historians, Darwin himself very much wanted a rapprochement of his theory with that of Lamarck; he wanted to include purpose in his theory. *In this book, I complete Darwin's program of including purpose in the theory of biological evolution, reconcile and integrate Darwinism and Lamarckism, define what makes us different from our animal ancestors, and chart humanity's future destiny.*

When all is said and done, you will see that although the current picture of human civilization is full of doom and gloom, the future of humanity is bright and full of promise. But we have to remember that consciousness works through us; a substantial fraction of humanity must join together in following what has been the message of consciousness all along: we need to transform from our base-level condition. It is not enough that we develop a worldview of meaning and purpose, we must live by meaning and purpose whose embodiment happens only through us.

Just to be clear, what is meant by transformation? If you consult the Internet, the many answers are bound to confuse you. Reminds me of a joke. A man—an avowed atheist—was sightseeing in the Yellow Stone National Park. In his enthusiasm, he must have gone a little out of the beaten path and suddenly a big bear confronted him. The atheist went out of his mind momentarily in fear for his life and cried, "Oh, my God, save me." There was so much sincerity in his call that everything stopped, and God responded, "You have been denying me all your life and now you want me to save you? Are you ready to transform? Become a believer?"

The atheist thought for a moment and replied. "No. I can't change. But maybe you make the bear transform from its violent ways." God said, "Ok. I will try."

The world started moving again. The bear ran toward the atheist but amazingly knelt down before him. The man was anticipating with great hope for his life. The bear said, "Thank you God for providing me with this food" and ate him.

Understand! Transformation is not a mere change in beliefs nor is it ritualistic. Transformation in the pursuit of meaning and purpose is spiritual transformation. In our ordinary ego, we are of constricted consciousness, we lack the presence of the embodied spirit within us. As we embody archetypal meaning and purpose, we live in spirit more and more in expanded consciousness. This is transformation.

Establishment Biology and Paradigm Shift

The current science of biology is incomplete in another way. It is mostly evidence-based; theory is lacking. The empirical data are not fully explained with fool-proof theory. The forte of current biology is engineering, handling and manipulating bits and pieces of biological matter to solve problems of living. In the absence of convincing theory, however, the bio-engineering manipulations—genetically altered food and RNA-manipulated vaccines are examples—have become controversial.

The poverty of the theoretical side is best exposed by the fact that current biology cannot answer the simple question What is life? It empirically defines the difference of living systems—the capacity for maintaining an integrity, the capacity for reproduction and evolution, the capacity for forming society— and nonliving systems that lack these capacities and then asserts existentialism: this is the way it is. Life appears when macromolecules of living—Proteins, DNA, and RNA—are somehow assembled in cytoplasm in a cell within a wall never mind that the probability of such a thing happening from random movements of matter is minuscule as many serious calculations demonstrate (Shapiro, 1989; Birch, 1999). Biology is chemistry according to this molecular biology—but it is a dogmatic belief. Its only virtue is that it fits nicely with the current dogma of majority scientists—everything is matter—scientific materialism.

It gets worse when we attempt to give a biological basis to psychology via neuroscience where the issue is consciousness. The current best materialist model of consciousness does not agree with neuroscience data. Story is that a neuroscientist had a bet on this matter with a quantum philosopher and the neuroscientist has conceded defeat. Details later.

Biologists within materialism tout Darwin's theory of evolution (or its revision Neo-Darwinism) as a definitive final theory. However, Darwin's theory is demonstratively incomplete on both theoretical and experimental aspects (Goswami, 2008; see also chapter 3-5). Most importantly, Darwin's theory is based on the fundamental necessity of organisms to survive. But where does the survival necessity come from? Molecules as defined in physics and chemistry do not have survival necessity. Neither is there any empirical evidence for the emergence of survivability in macro-molecules via material interactions alone.

It is the theoretical foundation of current biology—material-ism—that is wrong and misleading. The other popular ways to think of reality to which half of the world's population still believe in—religions—are also mired with dogma, mired with many incon-sistencies just like scientific materialism.

There is a third worldview—the spiritual worldview. The world is spiritual, it is made of consciousness. This is the worldview quan-tum physics leads to and quantum science is based on: conscious-ness (and its quantum possibilities) is the ground of all being.

The Judeo-Christian religions do not buy the idea of evolu-tion of life; but this spiritual ontology has no qualms about evo-lution. There is the yuga theory of evolution in Hindu puranas: humanity started with a golden age (*satya yuga*) but ever since it has been a descent and now, we are deep into *Kali yuga* of a dark age. However, an ascent may begin any time. A more recent theory of evolution based on the primacy of consciousness has also been proposed which is more scientific (Aurobindo, 1996; de Chardin, 1961).

The establishment scientists avoid the idea of primacy of con-sciousness like it is plague; it smacks of purpose and spiritual values which they reject. However, open-minded philosophers do consider the idea and present some legitimate questions. We must address these questions.

Can the Universe be Spiritual and yet Incorporate the Material?

We start with both why and how questions. The first question is of course, Is the universe *really* spiritual? Most of us experience our universe as a vast collection of galaxies, stars, and planets; of these, we know for sure that there is one little insignificant planet—our earth—that has evolved human beings who make philosophies about our being; but can these philosophies be more than mere musings and disturb the universe?

Behold! We have the capacity to experience the universe as separate from us. The vast majority of people, scientists who research these things included, miss the implication of what this means. They try to ignore the importance of *subject-object distinction* of manifest experience of the living and the sentient. I (subject) see a star (object); the subject-experiencer and the object-star belong to different categories of logic. We are not insignificant parts of the universe; we are of the same status or even of elevated status. Also, we can never make models of the subject—us—with objects of the universe.

And yet our scientists make models of life and its evolution on the basis of what they call scientific materialism—matter, objects explain everything. Obviously, their endeavor is fallacious.

The word 'spirit' today borders on controversy. If we agree that it really means a consciousness of oneness of everything, a oneness that transcends the subject-object split, then the concept of spiritual universe is an ontological statement: Consciousness is the ground of all being. This ontology is called monistic idealism.

A few people, a relatively small minority to be sure, have experienced fully what they call cosmic spirituality—an experience of oneness of everything, every object that we experience, and that includes the stars. However, all creative people hear the voice of the cosmic spirit via intuitions and creative insights. It is they who develop new ideas on the basis of their insights, it is their ideas that are helping us to build a science based on the primacy of consciousness.

It is a given for most scientists that biology should explain psychology. The easiest way to see that we need a paradigm shift in biology is to examine the theoretical situation in the field of psychology. Four forces have contributed to psychology: 1) psychoanalysis (Freud, 1961), Jungian psychology (Jung, 1971), and subsequent depth psychology—all based on the idea of two levels of the human psyche, unconscious and conscious. The idea of the unconscious has even been experimentally verified. 2) Cognitive-behavioral theories—our behavior is all conditioned and it comes from the brain. 3) Humanistic-existential psychology based on the above-mentioned philosophy of humanism; this force thrives on the idea of human creativity (Gruber, 1981; May, 1994). 4) Transpersonal psychology (Maslow, 1971; Grof, 1998; Wilber 1993) which proposes that there is a transpersonal self beyond our behavioral ego. This, too, has been verified by neuroscience data.

As you can see, establishment biology can provide a conceptual basis only for the cognitive-behavioral theories of psychology. The conclusion is straightforward: in order to incorporate the ideas of the unconscious, creativity, and transpersonal self, we must develop a new biology.

Attempts at developing new-paradigm biology have been made as well, albeit few. One such attempts has been the above-mentioned theory of spiritual evolution by Teilhard de Chardin and Sri Aurobindo. Another theory is that of the epigenetic and non-material morphogenetic field invoked for the explanation of cell differentiation—how the cells belonging to different organs of the human body have their genes activated differently in order to make different functional proteins (Sheldrake, 1981).

The discipline of medicine is another direct application of biology. Here again, there now exists theories of quantum processes involved in healing [Dossey (1982)—nonlocality, Chopra (1989), quantum leap], that have been verified by experimental data. The experimental verification of these theories directly demonstrates the inadequacy of the current Newtonian biology.

Ideas such as the unconscious (implying that there is a second domain of reality beyond that of what we are conscious—namely space and time), nonlocality, quantum leap, transpersonal self-identity, nonmaterial organizing fields, are all beyond the scope of Newtonian materialist biology. And yet all of these scientific theories have found experimental support to varied extent. However, if they seem like disparate ideas, they are; they need to be woven into an integral whole. This is what quantum science has been able to accomplish.

One other main source of confusion is this. Evolution of life initially is an evolution based on the survival motif; this has given us certain, mostly instinctual, aspects that I call the base-level human condition characterized by me-centeredness, negative emotional and pleasure circuits in the brain, and a tendency for processing unambiguous information rather than the ambiguities meaning brings. Materialist models of reality *are* seemingly adequate to explain this base-level human condition. If you think that is all there is, isn't it logical to assume that talk about spirit, spiritual values, transformation are all humbug, idealistic imaginations?

However, that base-level condition applies to roughly 85% of today's people (Hawkins, 2020). The ideas of spiritual science appeal to some 15% of humanity that looks for transformation of the base-level condition. And indeed, a substantial number actually manages to achieve transformation at least in part and their achievements *are* noticed. Recently, even the newly coronated king Charles of the United Kingdom came out with high praise for the societal contribution of these people, no kidding.

Customarily, scientists pay more attention to exceptions than to the rules because these exceptions lead them to newer and newer ideas of science. The current situation is peculiar because both spirituality (via simplification in the form of the religions) and science (via simplification in the form of scientific materialism) has become dogmas; *exceptions are ignored.* The two dogmatic worldviews—religion and materialist science—fight the game of power.

Materialists are stuck with the predict-and-control philosophy of determinism of Newtonian physics. This prejudice stops them cold when it comes to explaining life and sentience satisfactorily, when it comes to dealing with matters of purpose, meaning, and creativity, when it comes to people adopting values and transforming their human base-level condition. So, materialists do biology and psychology avoiding these hard questions.

The philosopher David Chalmers (1995) made news by challenging neuroscientists that they evade the hard question of conscious experience. Apparently, Chalmers had a bet with a neuroscientist and in view of the recent revelation that the computer science-based model of consciousness which predicts a brain circuit of consciousness does not agree with neuroscience data, no such brain circuit has been found; the neuroscientist has conceded losing the bet.

The idealist ontology breeds difficult hard questions as well: if consciousness is cosmic how come individual sentience? how is consciousness embodied in matter to produce individuality? If we are all one, how do we become separate? Undeniably most of us experience ourselves as individuals, as independent separate objects in material bodies that interact via sensory signals. At best, we may allow that we are part of a locally connected society, but oneness of everyone, everywhere?

The traditional promulgators of the spiritual universe say in response that the separateness we experience happens because of forgetfulness, a veil of illusory ignorance separates us from oneness, from Reality. Remove the veil, and you will be one with Oneness; and moreover, this expansion of consciousness will bring you bliss.

But this does not satisfy; it breeds why questions: if Oneness is bliss, why do we need to be separate? Why individual sentience? Why manifest life at all?

The idea of the spiritual universe is very old, could be as old as 7000 years by some estimates. So, people have been asking these how and why questions for millennia. The answers at best have been lame. In the olden days and even now, people call this Oneness by names like God, Ishwar, Allah, Eins Sof, etc. causing further confusion; is God like us humans, only super-duper, almighty? And the connoisseurs would say, look! God is ineffable! Huh?

Connoisseurs also say, the separateness is God's play, for fun. But, as we all know, more often than not, separateness is not fun, anything but fun, it brings loneliness and suffering.

The connoisseurs chuckle. Yeah, it sure seems like that. Learn to endure; be resilient, seek God earnestly. See the suffering as opportunity to seek a new way of living that will bring you happiness. Meditate and find out that really is bliss waiting at the end of your journey.

Maybe. But it is equally clear that these propounders of the idealist model of reality continuing all the way to transpersonal psychology also evade the hard question of how consciousness and its ideas can be embodied in matter.

The fact is that this mutual evasion of the hard questions will continue until we find a way to integrate the two basic worldviews, one based on primacy of matter—material realism—the other based on primacy of consciousness and ideas—monistic idealism. This would involve integrating dogma-free science and dogma-free spiritual wisdom traditions which is also the esoteric component of all religions. This would also entail the integration of all the disparate theories already developed and noted above. So why are we talking about this here? The answer in short is QUANTUM PHYSICS; it integrates science and spirituality; and it enables a quantum biology of purpose based on it incorporating Lamarck's and Sheldrake's important contributions. It connects transpersonal psychology, psychologies of the unconscious, and cognitive science. It gives us a unified scientific handle on matters of human sociology and civilization. Most importantly, it gives all of us humans a holistic life style to pursue that brings us health, prosperity, happiness, and intelligence.

Let's discuss quantum physics. Why is there so much controversy about its application to biology and psychology?

Quantum Physics and Its Paradoxes

You are likely to have been told that quantum physics has many puzzles and paradoxes that nobody can fully resolve. This is what prompted Richard Feynman to say, Nobody understands quantum mechanics.

The major paradox is this: quantum objects are both waves and particles. This is a logical paradox because waves can be in many places at the same time whereas particles can be only at one place at one time. Quantum mathematics helps; according to the mathematical equation of quantum physics, quantum objects are waves, unequivocally, with many possible positions. When we measure them however, they become particles. Quantum mathematics enables us to calculate only the probability as to where the particle will appear upon the collapse of the wave into particle.

There are only two legitimate possible ways this paradox can be resolved. The one favored by most scientists is called the statistical interpretation. The collapse is random and acausal; this is why only the probability of collapse to a particular position can be calculated. This has the advantage that the metaphysics that matter is everything does not have to be challenged.

But causality is part and parcel of science. In causal terms, these scientists are assuming random and (implicit) nonmaterial causes (see below) for the collapse which provoked Einstein's famous reaction: God does not play dice.

Also, this interpretation evades any answer for what happens for a single object. This is okay for physics and chemistry where we always deal with zillions of objects but not for biology and psychology where we have to deal with single objects—organisms. The materialist answer is feeble—there is no quantum biology or psychology. It is based on the assertion that at the macrolevel where living and sentient beings exist, quantum physics gives way to Newtonian deterministic physics. This assertion is dubious. There is vast amount of experimental evidence of quantum effects at the macrolevel of experience. Witness all the disparate theories to deal with them mentioned before.

The other way of resolving the wave-particle paradox is to adapt the spiritual ontology—consciousness is the ground of all being—in the language of quantum physics. Quantum objects are objects (waves) of possibility within consciousness in the domain of potentiality for consciousness to choose from. When consciousness via an observer/experimenter chooses/measures, the possibility wave collapses and becomes a particle in the manifest experience of the observer.

Consciousness and the waves of possibility within it reside in a domain outside of space and time, the domain of potentiality, akin to Freud's conceptualization of the unconscious in psychology. Although quantum measurement involves an observer, it really is a choice from Oneness that collapses a quantum object's possibility wave. Finally, in the process of the quantum measurement, choice, and collapse, consciousness identifies itself with the observer's brain creating a distinction between self and the object in the experience of the observer (Goswami, 1989, 1993).

Observe! Collapse is not just collapse of the object's wave aspect into its particle aspect; collapse is a product of choice and intention for distinction; the effect of collapse is a split of oneness into a clear distinction of a subject and an object.

In the book *The Self-Aware Universe*, I solved the problem of how the embodiment of consciousness in matter takes place in this quantum science of consciousness; in this way I removed the major argument against the theory of a spiritual universe. In this book, I will demonstrate how the ideas of consciousness can be embodied in matter; I will formulate a new theory of how quantum physics can apply to the macrolevel of experience in spite of decoherence. Actually, sneak previews of this latter have already been published (Goswami, 2024; Goswami and Onisor, 2024).

Note. Quantum movement in submicroscopic objects is coherent: all the components dance in phase such as in a chorus line; in macroobjects, the different quantum components fall out of coherence, they dance rock and roll. This is called decoherence.

Quantum textbooks usually mention a third compromise solution of the wave particle paradox called the Copenhagen Interpretation fathered by the physicist Niels Bohr. In Bohr's interpretation the wave and particle are two complementary aspects of the quantum object; when the human observer measures, he chooses to measure either of the two aspects. The collapse event itself is acausal; no cause can be given of it.

But the complementarity principle that Bohr uses to hide the domain of potentiality is demonstratively false. Both particle and wave measuring arrangements give away the secret of the solution of the wave-particle duality—that you never see the wave aspect in manifestation, that the wave belongs to a different domain of reality outside space and time—upon clear analysis. Read my book *The Everything Answer Book* for details.

We will explain the details later in the chapter about how the embodiment of consciousness in the brain and the subject-object split happens; but this means quantum physics is giving us a scientific answer to that age-old inquiry: how does the oneness split into subject-object separateness that we experience?

How about the skeptical inquiry—if Oneness is so great—the abode of bliss, spiritual masters say—then why create separateness, ignorance, and suffering? The answer is that the play of conscious manifestation is not a frivolous play but a play that has purpose; in other words, the purpose of manifestation—to embody itself as well as all its qualities and capacities—is to bring love, beauty, truth, justice, wholeness etc. in human living. Teilhard said it best, the purpose is to bring heaven on earth.

Also, quantum measurement theory is forcing us to the realization that consciousness as the ground of being with no subject-object split awareness is the domain of potentiality *outside* space and time. In this concept, we have found a scientific explanation of the Freudian and Jungian concepts of the unconscious which were originally based on empirical data.

Why is there controversy about applying quantum physics to our life sciences? Like the paradoxes, the controversies are also created by trying to maintain the grip of scientific materialism over science. Clearly absurd and experimentally unverifiable solutions of the quantum measurement problem such as the many worlds theory are proposed as alternative to the idealist solution above that enable consciousness to enter science.

As the great physicist Paul Dirac said, great advances in science require giving up great prejudices. In order to solve biology's unsolved problems, biologists have to give up matter moving in one domain of space time philosophy and adopt the two-domain metaphysics for science based on the primacy of consciousness.

Psychophysical Parallelism

In quantum science, a breakthrough mechanism that makes the embodiment of ideas of consciousness in matter possible is called *psychophysical parallelism* first conceived by the philosopher Gottfried Leibniz. Matter and the organizing principles (blueprints) of life are both quantum possibilities of consciousness to choose from. As consciousness chooses actuality from possibilities of the organizing field, it also chooses appropriate function of an appropriate material organ in the living body, that is, the organ's physiology. At once the organism has an experience of the movement of the organizing field that is mapped into the physical body as an organ function. For example, the idea of survival comes to us via the organizing field for cell differentiation in organs (see later for further elaboration) and is implanted in the physical body as the physiological function of digestion in our navel organs (fig. 1).

Is there Quantum Movement in Cellular Matter Participating in Living?

This book is also about quantumizing our current ideas of biology so that biology can provide the long-expected connection between quantum physics and the psychology of consciousness. We will see that the arenas of these traditional scientific fields of biology and psychology are the result of progressive embodiment of consciousness and its attributes in matter that happens during evolution, first of animals and then of humans.

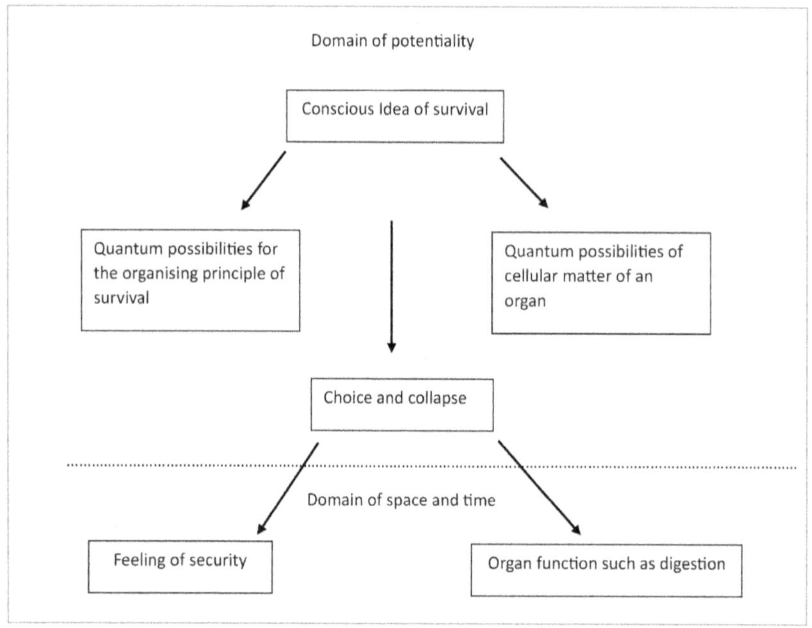

Fig.1. How consciousness creates psychophysical parallelism

The physicist Erwin Schrodinger wrote a book *What is Life?* in which he wondered about this question: if living cells are made of atoms and molecules whose movement is quantum, how could living systems of cells be stable? He was thinking of the uncertainty principle—principle of indeterminacy. The conclusion he reached was simple: living matter—molecules important for cellular function such as DNA, RNA, protein etc.—must be big enough for quantum movements of uncertainty to give way to Newtonian deterministic movement. But still one thing never stopped puzzling him: genes—portions of DNA—have the program to make the important cellular functional molecule of proteins; but genes are of a size of roughly three hundred thousandth of a millimeter and would contain "no more than about a million or a few million atoms." That he knew was too small a number to turn off the uncertainties of quantum movement.

If you have never heard of the uncertainty principle, a joke will illustrate it perfectly. The co-discoverer of quantum physics Werner Heisenberg was driving in Berlin when a policeman stopped him. "Do you know how fast you were going?" The policeman inquired. "Where was I when you zapped me?" Heisenberg asked back. "You were right at the center of downtown Berlin," said the policeman. "Oh, drats. Now I do not know, cannot know how fast I was going," said Heisenberg.

Got it? According to the uncertainty principle, we cannot simultaneously measure perfectly accurate values of both position and velocity of an object.

Coming back to genes, indeed, later the physicist Walter Elsasser (1981) asserted that gene mutations are indeed quantum. And here we need to assert that the quantum cellular matter referred to in fig. 1 (I call them quantum gene activators, see chapter 7) also retain the uncertainties of quantum movement.

In other words, evolution and development (Evo-devo) of life are not evolution of Newtonian matter as scientific materialists and traditional biologists think. Instead, evo-devo is the story of consciousness guiding quantum cellular matter to progressively make better and better embodiment of consciousness and its experiences including the experience of its attributes—the archetypes—in matter.

Even a cursory look at the level of consciousness that most people live demonstrates that we are far from being done with archetypal embodiment, we are a project needing further development. The good news is that with a better theory today, we know the creative process needed for transformation. We can also predict the future with some confidence. And the predicted future is wonderful as you will see. The problem is how to go there from here. How to find the motivation to transform ourselves and our society?

We Can Change but We Need to have faith in our Ability to Change

Traditional biologists, behaviorists, and cognitivists say we cannot change our animal nature, our survival orientation. The humanists and transpersonal psychologists have already called their bluff based on philosophy and experience. Now quantum science is backing the latter up with solid scientific theory.

The latest good news is this and this one is from neuroscience. Can you make changes? Neuroscience data is saying yes. You can, provided you change your mindset, your worldview.

Psychologist David Hawkins did some measurements using muscle testing that showed that most people—maybe 85%—fall in the **No-can-Change** category; they would rather stay where they are—the base level condition. However, roughly 15% of humanity have already embraced a **can-Change** mindset.

Stanford researcher Carol Dweck was researching on people's achievement and success. She discovered the same thing that people subscribe to one of two mindsets.

Fixed mindset: The belief that what you are born with is what you got. If the base-level human condition is it, then that's your lot.

Growth mindset: The belief that you can change, you are not stuck with what you are born with. You have in potentiality the innate creative ability to scale the ladder of success, happiness, and intelligence. You can learn, improve, and transform to a new you by developing your creative abilities. Different names for the mindsets but the same idea.

Most people of the fixed mindset accept the so-called experts' view of reality that the mainstream media propagates—scientific materialism. As they struggle with success, they want to quit, and the worldview backs up their decision. Why battle with failure? Why not embrace "I can't change" fixity – and accept that you are what you are: not creative, not intelligent, not adept, not good enough, etc.?

People who embrace an "I can change" mindset on the other hand develop their creative potentiality, accept failure as temporary setback, develop resilience. And "let the world be their oyster."

You know about neuroplasticity, right? There is neuroscience evidence (Moser et al, 2011,) that as you accept a "can do" mindset, your brain develops what is called error positivity wave response which reflects your new ability of paying more attention to your mistakes and make failure a learning opportunity.

You are no longer afraid of making mistakes; you'd fail rather than do nothing.

Another neuroscience study (Myers et al, 2016) found that a growth mindset leads to new brain wiring connecting brain areas improving error monitoring and learning from mistakes.

This is reassuring suggesting that adopting a can change mindset, you can embody a major aspect of creative behavior.

The best part of all this development is that this new paradigm of quantum science makes it clear from the outset the following: The materialists are not totally wrong; they are talking about the ordinary human consciousness which *is* almost all mechanical (the only real freedom left is the freedom to say "no" to conditioned behavior). Quantum physics enables us to see that the mystics and humanists and transpersonalists are talking about potentiality (remember: "quantum physics is the physics of possibilities") that is available to everyone but you have to actualize it to experience it and then practice living it to embody it, and all that takes engaging the brain, actually your life itself, in the creative process of transformation. The missing link between the two approaches is a comprehensive theory of life and the brain and a theory of creative transformation. Preliminary work on all this has been done; read my books, *Creative Evolution, Quantum Creativity*, and *The Quantum Brain*, the last co-authored with Valentina Onisor. And now science and spirituality both can be accepted as part of a developing human quantum science of everything that is living and that is human.

In this spirit, let's expand quantum physics to quantum science that includes and explains all our experiences. In the meantime, please remember *how not to think about a human being, life, and consciousness:*

- Living beings are not merely bundles of molecules.

- Survival is not the only purpose of living.

- Consciousness is not energy.

- Consciousness is not a hologram.

- Consciousness is not a carbon computer with cognitive ability.

- Consciousness is not a product of complex information systems.

- Consciousness is not a brain phenomenon coming about due to complexity.

- Consciousness can be conceptualized as the one and only quantum field—the ground of all being—*with both subject and object potentiality.*

The one idea to take with you:

> *The theories of quantum science, nay, all scientific valid theories, are discovered by creativity, going beyond rational thinking. Scientific theories are not mere philosophy; they literally are experimental philosophy. Their validity can be and must be verified by experimental data.*

Back to Evolution and the Ascent of Humanity

It is likely that all life on our planet, including human life, originated from a single cell; the exact date of the origin goes back to 3.8 billion years ago although more recent data is suggesting that it could even be earlier, 4.2 billion years ago. And ever since we are evolving. Evolution data, some controversy notwithstanding, is pretty conclusive about that. But then questions raise their head whose answers are hard to find.

Two of these questions go like this: 1) are humans the pinnacle of evolution? 2) Are they going to ascend further? In this book, I am going to establish that according to the newest science—quantum science of consciousness—the answer to both questions is yes. Why is this important? *You, we, humanity need to participate in the evolution process; else we cannot progress. I will demonstrate that big evolutionary changes have always come from group or community consciousness and creativity, only when large communities of a species joined in the creative effort. The new science is showing us the scope of our participation and chalking out some details.*

The fossil data is quite definitive about some sort of ascent; it is the outdated but politically correct theory of Neo-Darwinism that keeps the scientific establishment confused. The fossil data does conclusively suggest that evolution has proceeded from simple expressions of consciousness to complex expressions of consciousness, something like an arrow indicating which way the time flows—that is, a *biological arrow of time is real* (Brough, 1958). In the beginning all you find are simple fossils consisting of single or at best multiple cells loosely put together. The experiences of such creatures likewise are highly limited. Then come creatures with cell assemblies called organs which perform specific purposive functions suggesting greater scope for conscious expressions. Eventually, organisms with the brain evolves and the capacity for conscious self-identity.

For a long time, arguably, the purpose served was simply survival. This one purpose seems to be adequate to explain most biological data on evolution from amoeba to plants to animals. This created confusion enough for materialist biologists to reframe Darwin's theory in a materialist scenario that seems to have genuine explanatory power. Implicitly, the assumption was that survival does not count as a purpose; instead, it provides the genes a causal incentive to evolve—the concept of the selfish gene (Dawkins, 1976).

Eventually, molecular biology officially declared that biology is chemistry; survival is an emergent property of the complex macromolecules of life. Unfortunately for the protagonists, there has been no experimental verification of that idea in the laboratory (Goswami, 2008).

In truth, with the evolution of the brain, especially of the cerebral cortex, it is clear that evolution, besides survival, also serves what can only be called *higher purpose*. With the help of the cortical brain humans are able to process meaning of the objects of their experiences with a new attribute that we call the mind including objects of intuition called archetypes that define purposive values—truth, love, beauty, justice, wholeness and all that—for us.

And more. As we explore our latest evolutionary organ—the cortical brain—we find that there is room for opening further doors. The evolutionary data clearly shows that the brain of the humans is already the most versatile in terms of functions performed and its functional potentialities have hardly been explored yet.

Darwinism or Neo-Darwinism, not having the concept of biological arrow of time within it, not having incorporated a theory of development within it, has no comment to make about the future of evolution. Naturally, most biologists profess no interest in the question of if evolution is the story of the ascent of humanity to the pinnacle or about the future of evolution.

But if you don't belong to the community of professional biologists, What then? After you read this book, you should have no doubt that humans are the pinnacle of evolution and that they should ascend further.

Earlier I said that closer examination reveals that molecular biologists bypass the question *What is life?* rather than try to explain it. In this way, the biological establishment can pretend there is no problem and go on enforcing molecular biology's validity over the entire biological community.

But you can fool some of the people some of the time but not everyone all the time. Paleontologists you know; they are the scientists who gather data about fossils (you probably have heard the joke: why do paleontologists fall in love with fossils? Because they do carbon dating). Recently, they have dug up some new data that contradicts establishment ideas about evolution. In 2016-17, there was a big debate on the Internet about the new data, about evolution in general, and also about the question, What is Life? among both materialist and post-materialist thinkers.

If not molecular biology, what other scientific theory should we use to guide us to develop an evolution theory beyond Neo-Darwinism that will give us answers about our past and about our destiny? Unfortunately, the debaters could not reach any consensus regarding either of the two questions.

Fortunately, these questions have been satisfactorily addressed in my book *Creative Evolution* published in 2008, and even before that in two scientific papers published in the nineteen nineties (Goswami, 1996, 1997).

These questions raised in those discussions and quantum science solutions are all the more important now, because if you look around in an unbiased manner, human civilization is decaying now, not ascending, all the technological progress notwithstanding.

Maybe we are in the dark age of Kali yuga after all! Can we arrest this descent and get human civilization and evolution on track, ascending again?

The specialness of the human being is that with its big brain, a human can bring forth its mind to give meaning to all its experiences. As hunters and gatherers, we gave meaning mostly to our physical experiences. This was followed by the garden agricultural era when we gave meaning to mostly our feelings and developed *new positive emotions. Then came the large-scale agricultural era and rational* mind—mind giving meaning to meaning itself. Most of us are still doing that although a few of us have already moved on to the next phase of human development—giving meaning to intuition.

As previously noted, the ascent of civilization is measured in terms of the increase in the scope of expression of consciousness.

From self-domestication to altruism to the development of positive emotions and the collective unconscious, the scope of conscious expressions has only gone up and up.

At first sight, it seems that the advent of the era of the rational mind should only help further embodiment of positive emotion as more capacities of processing meaning are added. Indeed, it did initially; witness for example the Vedic society in India in its early stages which was egalitarian. Unfortunately, rational thinking is step by step, algorithmic, it favors a simple hierarchy. Moreover, with the development of the idea of private property, there was added interest of innovative leaders toward exploring the archetype of abundance. To mollify greed associated with the exploration of abundance, exploration of goodness and love are not enough; the exploration of higher purpose, higher archetypes like truth, justice, and wholeness are needed, and that did not happen, not enough anyway. In the rational era, societies became simple hierarchical as a result. People did not participate in shaping society anymore; the elites—kings and the religious oligarchy—guided the society for better or worse. The exploration of higher archetypes was confined to a small middle class of professionals.

With the advent of modern science, we have been making new efforts to increase people's participation in determining the movement of the society. Witness the rise of democracy and capitalism. But these systems worked only so far as the leaders were willing to mollify their exploration of power and abundance with ethics and morality. When the archetypes were banished via the rise of scientific materialism, democracy and capitalism began to serve the elites once again. This is the situation right now.

In the past, the elitists were loosely united in their worldview; with the advent of scientific materialism, the new elite of meritocracy became oppositional to the old elite of aristocracy and religious oligarchy. The descent of civilization going on now is a sign of rampant powers struggle between the new and old elitists.

The fact that that we are approaching a crisis that threatens our survival indicates that time has come to integrate all the disparate worldviews and factions once again, and this is precisely what is happening with the paradigm shift in science. A future ascent from rational mind to the intuitive mind will follow. As they say, when winter comes, can spring be far behind? There are signs that the number of people of intuitive mind is growing; it is already as large as perhaps 15%.

There is already sign that archetypes are returning to our lives. The researcher Sara Konrath did a study of young people's altruism (she labels it inaccurately as empathy) in 2011 only to find it that it has plummeted compared to the peak it reached in the 1970's. However, at this writing (2024), she has published a follow-up report which shows that altruism is back up to the 70's level.

With the current crisis facing us and quantum science leading us to regain faith in ourselves, we already have motivation to change. All we need is leaders with spiritual transformation and moral authority who walk their talk. And this is already happening.

I am optimistic about our future. I hope that after reading the book you will be too.

Take with you a few important things to remember if you want to join the movement toward the intuitive mind using the new integrative quantum paradigm and worldview:

Science is not materialist science; science by definition cannot have unverifiable dogma.

Spirituality is not religion; spirituality—Oneness of consciousness—is not dogma, it is an integral aspect of the human experience; anybody can experience it with some effort; just love someone deeply and see.

Once you wake up to the truth of Oneness, all archetypes are categorically imperative for you to explore and embody as the philosopher Immanuel Kant intuited long ago about the archetype of goodness; you will never find full satisfaction until you do.

Our society has to find ways to establish respect for the archetypes. Only with people providing leadership of moral authority that can happen.

The one thing I am sure about is this. There are two societal movements going on. One is descent with the advent of materialism in a scientific form—materialist science—and the reaction of religions to it. The other is ascent led by people dedicated to transformation—their clients are about 15% of the population. This second group needs to make a difference. They have to turn things around. First, unite with one front of a new paradigm of post-materialist science. Quantum science gives you the only promising scientific front right now with wide enough scope to guide people in dealing with all their experiences and thus influence people's worldview. Only quantum science is able to integrate physics, biology, psychology, health sciences, and even social sciences.

Let's rally around the new paradigm of quantum science; this is not the time for idle philosophies such as holism or panpsychism, we need transformation of self and society, real results. Let's aim high and be leaders of the upcoming era of the intuitive mind; let's transform ourselves and change society while we walk our talk: this is what we call *quantum activism*. Read the book, *How Quantum Activism can Save Civilization* for further information.

Is there a way out of the decline? There is. But we the 15% have to take coordinated action.

Creating Positive Emotional Brain Circuits and Making them Universal

I have been arguing for some time, most forcefully in two books written in collaboration of the integrative physician Valentina

Onisor called *The Quantum Brain* and *The Awakening of Higher Intelligence* about rewiring the brain for further transformation toward intelligence—morality, happiness, and enlightenment. This requires us to develop positive emotional brain circuits. The challenge is to find motivation.

The alternative is grim, that is one motivation to begin with. Our worthy ancestors have already charted the path; it is called hero's journey. It consists of exploring and embodying the archetype of power in conjunction with the archetypes of love and goodness, the same recipe we need for exploring both archetypes of power and abundance and mitigating them with not only the archetype of goodness and love but also with archetypes of truth, justice, and wholeness. What this amount to is making positive emotional brain circuits enough to balance the built-in evil in us in the form of the negative emotional brain circuits.

For the 15%, this is the easy part. The difficult part is how to get some respect and support for transformation from the other 85% who live more or less on *eat, drink, be merry* permissive lifestyle. Today, they are dumbed down having cluttered their brain with useless information and are easily manipulated by elitists.

We have to somehow make the positive emotional brain circuits universal like the negative ones. That is, to make changes in our collective unconscious so that all people will be born with the ability of being good to everyone and balance their negative tendencies. Is that possible? I think the answer is yes (see chapter 9).

The purpose of the book can now be succinctly summarized:

I demonstrate molecular biology's errors and develop quantum biology to take its place.

I use quantum science to integrate Aurobindo and de Chardin's basic ideas with those of Lamarck and Darwin to formulate a theory of evolution-development (quantum evo-devo) that demonstrates the ascent of the embodiment of consciousness in living beings. I also present empirical data in support of quantum evo-devo.

I also present the processes involved in the ascent, and also why these processes have currently stopped. I then present some ideas of how to initiate a restart of an ascendant evo-devo once again.

I argue that what is needed for a restart is group participation in a transformative creative program for directed evolution.

Can we initiate the restart toward another golden age? I am convinced that if the 15% I keep mentioning put their intention and creativity in this most ambitious endeavor ever conceived, then we can do it. Are you interested? Are you motivated? I hope you already are and will be further motivated by this book's message.

A final note. Quantum science shows that following the hero's journey expands our consciousness; it is bliss. Doing it as a group will further enhance the bliss. As the mythologist Joseph Campbell used to say, Follow your bliss. That should provide the needed additional motivation. For those of us who can, let's do it. In the process of exploring archetypal bliss, let's take humanity to its predicted future ascent.

CHAPTER 2

An Intro to Quantum Science for Nonspecialists

According to quantum physics, material objects are waves; they become particle only upon measurement. To appreciate the puzzle of this, realize that waves are objects that spread out like water waves; they reach many different places all at the same time. Particles, on the other hand, moving or stationary, always remain localized.

How can matter or energy be that way, behave wavelike? They can be if the waves are waves of possibility residing in a domain outside of space and time.

For example, as a wave of possibility, an electron can exist in many different positions, all at the same time, but in possibility. In contrast, when the electron wave becomes an actuality, a particle in space and time, it occupies only one position at a time, just as a particle must.

Possibilities are not part of the space-time reality where everything is concrete. Possibility waves reside in a separate domain of reality, the domain of potentiality. What differentiates this domain from space and time?

In space and time, objects communicate via signals moving from one to the other. Signals have a speed limit, the speed of light. So, communication through space and time, called local communication, always takes a finite time.

In contrast, communication between objects in the domain of potentiality can be instantaneous if the objects interact and achieve a special state called state of correlation or entanglement. How can that be since signals have a finite speed? It can if communication between correlated objects is signal-less; then it would be instantaneous. We call this instantaneous signal-less communication nonlocal communication.

Now think. When do you experience instant communication? Only when you communicate with yourself, right? So, if all objects can potentially communicate instantly in the domain of potentiality, all must be part of a potential Oneness—consciousness in suchness.

What defines a measurement? The presence of a human observer. What defines an observer? A human's ability to experience himself as separate from the world of objects, an ability or quality we identify as the subject or self of an experience. In other words, quantum measurement must include the observer's brain whose job is to produce the subject-object distinction when quantum possibilities are collapsed into actuality.

Let us understand why an observer is necessary for quantum measurement following the mathematician John von Neumann. The way you see a measurement is likely this: what we call a measuring instrument does it, and we human observers just verify the instrument's work. But how do you think of the measuring apparatus such as a Geiger counter which goes ticking when an electron impinges upon it? Elementary particles make atoms, atoms make molecules, molecules make the Geiger counter. Now think the quantum way. Elementary particles and their assemblies are possibilities, right? So, possible elementary particles make possible atoms, make possible molecules, make a possible Geiger counter. Can a possibility of a Geiger counter coupled to a possibility of an electron produce anything actual? There is a problem! Von Neumann pointed out the obvious: no. He did even better. He took the mathematics of quantum physics and proved a mathematical theorem: no material interactions can convert possibility into actuality; possibilities interacting with possibility only produce more possibility.

So, what we call a measurement apparatus such as a Geiger counter cannot accomplish measurement or collapse, but merely amplifies the signal to facilitate the observer's perception. Even the human brain must be some kind of an amplifying apparatus since it cannot convert possibility into actuality by itself as a material object either. But nevertheless, it is undeniable that a human observer does hear the tick of the Geiger counter, does complete the measurement—the conversion of possibility into actuality. Conclusion: observer is more than the material brain; it must have a nonmaterial non-object aspect that is outside of the jurisdiction of quantum mathematics that applies only to objects.

This non-object aspect von Neumann saw as the observer's consciousness is what we normally call the subject or a self. He proposed that in a quantum measurement it is the observer's consciousness—the self—that "collapses" the electron's possibility wave of many possible positions into a particle of one unique position by choosing the one facet out of the many facets. But upon scrutiny, physicists found this idea paradoxical as well. The most famous is the paradox called the paradox of Wigner's friend: in the case of two observers for the same quantum measurement, who gets to choose?

For example, for the electron, suppose two experimenters have set up an array of Geiger counters in all the possible positions. They both are measuring the same electron simultaneously and each chooses the electron's position to manifest differently. Whose choice counts?

John von Neumann did not live to see quantum nonlocality become experimental fact in 1982 thanks to the experiment of Alain Aspect et al (1982) in France who demonstrated that correlated quantum objects do communicate nonlocally. Correlated quantum objects become one; there is a potential oneness that pervades the domain of potentiality.

So, what happens in a quantum measurement? A causal process from the nonlocal Oneness (of the domain of potentiality) consisting of choice from among the different facets of possibility of a possibility wave collapses the wave to a particle. Simultaneously, the Oneness identifies with the any and all observer's brains and becomes a subject (the quantum self). Since it is oneness that chooses, the position chosen is the same for all the observers.

In this way, the Oneness of the domain of potentiality is con-sciousness; it transcends both subject and the object. *Quantum measurement is not only a conversion of an object's possibility wave to particle/actuality via choice but also a conversion of non-dual one consciousness into a subject-object duo of distinction.*

The higher power of spiritual/religious traditions—downward causation—is seen in quantum science as the causal power of Oneness consciousness to choose and manifest in subject-object distinction.

In the process of measurement, observer's brain makes memory; repeated reflection in the mirror of memory eventually produces conditioning, a tendency for consciousness to choose the response chosen and reinforced before. So, there is the initial subject or self of the experience with the clear awareness of Oneness, I call it the quantum self; upon reflections in the mirror of memory, it becomes jaded as the conditioned subject of ordinary experiences that we call the ego.

Two predictions of the theory must be emphasized: 1) Human consciousness as we experience with the brain can be nonlocal (in the quantum-self mode); therefore, brain-to-brain direct commu-nication without signals should be possible. The experimental evi-dence is called transferred potential (Grinberg et al, 1994, Standish et al, 2004): brain electrical activity can be transferred from one distant observer-brain to another correlated by the two observers' ongoing conscious intention without signals (read my book *The Everything Answer Book* for details). And 2) us humans have two modes of the self, a nonlocal quantum self (the scientific name of the spirit manifest within us), and a local mode, the ego. This prediction has also been verified by neuroscience data.

There is now a ground-breaking technique for brain research called functional magnetic resonance imaging, abbreviated as fMRI. MRI you know; it consists of making pictures of the internal organs of the brain using the phenomenon of magnetic resonance. fMRI simultaneously measures the oxygenation, blood flow, to ascertain which brain parts are actively functioning while we take the image.

fMRI measurements clearly show that when we do ordinary tasks, the brain acts locally, alternating between a task-related area and a default area called the self-agency area. This is when we experience ourselves as the ego.

But then in a creative experience which long term meditators meditating on loving kindness experience quite frequently, the brain becomes nonlocal: different even distant parts of it become active with synchrony like a symphony orchestra. For the latter, the conductor's baton maintains the synchrony of the music of the individual players. For the brain, it must be the nonlocality of the quantum self. (Read my book with Valentina Onisor, *The Quantum Brain*, for further details.)

Things to keep with you:

Conscious choice is the new name of downward causation of the old traditions.

Quantum objects can communicate in two ways, a) local, with the exchange of signals; this they can always do; b) nonlocal, without signals; this they can do only when correlated.

The Oneness (nonlocality) of consciousness is now a scientific idea with both theory (quantum physics) and experimental data (of Aspect and collaborators) to support it.

The theory of quantum science gives two predictions both of which has been verified experimentally; a) any person's brain has quantum aspects via which consciousness can connect with it; correlated quantum brains of two persons can communicate nonlocally; b) human beings have two modalities of the self, the quantum self and the ego.

Does Quantum Physics Apply to Us at the Macrolevel?

Aspect et al's demonstration of quantum nonlocality creates a gaping hole in the belief system of scientific materialism. How can we claim that there is only one domain of reality when correlated quantum objects can communicate instantaneously using a domain outside of space and time?

Materialists try a tame response: We sentient beings have big bodies and live in the macroworld at room temperature. Here quantum physics gives way to Newtonian physics; the nonlocal avenue of communication ceases to exist; we no longer can access the domain of potentiality. For all practical purpose, we live in one domain of reality—space and time.

Recent demonstration that macroassemblies of quantum objects tend to decohere, that is, coherent quantum movement is no longer sustainable for them, especially at room temperature, further augments the idea that quantum physics does not apply to our brains, to us.

The approach of post-materialist research has been to try to show that somehow the human brain *is* quantum (Walker, 1971; Hameroff and Penrose, 2013) and then researchers try to explain brain's acquiring consciousness on the basis of the quantum brain. Hameroff and Penrose even bring gravity in the picture to give their theory credibility. Unfortunately, these approaches to consciousness are doomed from the beginning: if your model only has objects and their interaction to begin with, you are never going to get a subject/experiencer out of it: subjects and objects belong to different logical categories.

The approach of quantum science is different, through quantum measurement theory and the observer effect. In order to observe submicroscopic quantum objects, an observer has to employ devices like the Geiger counter or a photographic arrangement. These devices are usually called measuring devices; but thanks to von Neuman's work, we now recognize them as amplifying devices; they amplify the submicroscopic signal so a human observer can see it.

The big question now is, How does the observer see an object? Here there are philosophers called identity theorists who claim that the brain sees the object directly. Researchers who make computer models of consciousness see this direct "seeing" as the action of something like a CPU in the brain which then delegates finer aspects of refinement of cognition to the rest of the brain apparatuses. This is the model that now has been toppled by neuroscience measurements. There is no doubt about it, the multitude of brain apparatuses process the signal first.

In quantum science, consciousness is the ground of being in which the object, the amplifying apparatus, and the brain exit as decohered quantum possibilities. In the case of an experience of an external object, *the brain becomes momentarily quantum* and consciousness uses the quantum brain so that upon collapsing possibilities into actuality it can see the object separate from itself—for making distinction. In this way, the tendency to decohere is overcome for the brain only momentarily. This what is achieved by the observer's attention.

Think. Our brain is bombarded by external stimuli all the time most of which we are not aware of. Only when we pay attention, does the brain become quantum and coherent and consciousness can connect and use the brain for creating distinction (see below for more details) so we can observe and be aware of it.

The Quantum Science of Living Experiences: Sensing, Feeling, Meaning, and Purpose

Let's change the subject slightly, let's come back to the discussion of living. Living is about sensing, feeling, thinking, and intuition, the four experiences that occupy human beings most of the time. But of course, we don't all experience all of these things equitably. In fact, the psychologist Carl Jung did a study of human personalities based on which he divided people in four personality types: sensing, feeling, thinking, and intuiting. Today, most people probably would put themselves in the sensing and thinking category. Feelings come to us mixed with thoughts most of the time, a mixture we call emotions, most of them are negative and brain-based (the brain memory is five times more negative than positive, say neuroscientists). Traditionally, women are supposed to value emotions, and men value thinking; but this is rapidly changing. In today's culture, even many women are offended if they are called emotional.

Today, most young people would rather interact with others on the cell phone. I read the daily comics in the newspaper and in one of the comic strips, a young fellow is speaking to the shopkeeper on his cell phone putting his order in. The shopkeeper says, "yes we take mobile orders, but if you are actually inside the shop, we prefer that you speak to us." The cartoon is funny because everybody recognizes that this is the trend of the culture, but it probably has not gone that far yet. I always get a chuckle when I see teenagers walking together and speaking to each other on their cell. And that reminds me of another cartoon on the same pages of the daily newspaper. This comic strip is famous, called *Garfield*. Garfield is a cat; he is on a date. The girl cat says, "Are you going to stare at your phone through our entire date?" Garfield says, "Yes, I am." The girl cat says, "Cool," and starts staring at her own phone.

And if intuition seems foreign to you, you probably do not even know about it, know that you have this capacity.

And then if I ask, what do we feel, what do we think, what do we intuit, today's young people may very well be a little puzzled. Perhaps nobody asks them such questions.

What do you think? Most young people would say, "Oh, stuff, about people and things." Such "stuff"—thoughts without ambiguity—is formally called information. Information that you can put in your laptop; computers process information by mapping them as symbols. And today's young people mostly process information on the Internet in all their spare time. That is what social media is all about.

Computers process information. Are people computers in human form, robots? Many brain scientists say openly, yes, we are robots, robots with experiences. They label people as p-robots; *p* stands for philosophical. Because although we humans are robots in their estimation, obviously since we have experiences and robots don't, we are a different kind of robots. But the difference is only philosophical with no causal effect. Like regular robots, we are still mechanical, totally predictable behaviorally.

Why do these scientists concede that robots do not have experience? This is because an experience has two poles—subject and object, the experiencer and the experienced. We make robots from objects in our laboratory; we have no idea how to program a subject into the robot, because obviously the subject belongs to a different category of logic altogether to be a separate pole. In other words, it is impossible that we can ever make a subject, the experiencer out of the objects that the subject experiences.

Since you are reading this book, I guess you do not think of yourself as a p-robot; in fact, you likely *feel* that thinking this way is abominable and you are looking for a different depiction of the human being, one that is capable of handling the ambiguity of thoughts of meaning and purpose. You *know* in your heart that it is feeling, meaning-thinking, and intuition that define you, not robotics and information-processing. And you know also that *you* exist with causal potency. You are an experiencer— the subject-pole of experiences—with freedom to choose.

And more. You are not to be denigrated as a mere conglomeration of objects; your causal efficacy extends beyond the objects like brain-memory you are represented by; you are capable of creativity.

But of course, the propounders of the robotic philosophy, scientific materialists and rational thinkers have a lot of arguments against your view of yourself, against the ideas of human free will and creativity and these arguments are quite convincing to a lot of people, at least confusing enough to not worry about matters of philosophy. "Questions of meaning, purpose, self, free will, creativity? Those are questions for the philosopher. Why bother me?" Too many people think this way that the status quo that I call intellectual imperialism of materialist/rational thinkers—meritocracy—perpetuates.

Science, when it was free from dogma, was called natural philosophy. The scientific materialists did an about turn on this. Too much worrying about difficult philosophical questions can hinder progress. In that spirit, in the nineteen fifties, these scientists gave up on finding satisfactory answers to philosophical questions like, What is life? What is subject or self? What is the meaning and purpose of human life? How can mere molecules strive to survive? and just flatly declared, all is matter and material interactions. Enough already of philosophical debates about vital energy that we can never measure, about purpose for which there seemed to be no way to include it in science, Newton could not do it nor could Darwin or Einstein, about free will which only leads to religious bigotry about God and God's will, or about the unconscious which calls for a second domain of reality.

In truth the scientists of this ilk have gotten rid of God a long time ago. The eighteenth-century physicist of Napoleon's time Pierre Simon de Laplace wrote the first scientific book that did not mention God, not even once. When Napoleon asked him about it, Laplace said: "Your majesty. I have not needed that particular hypothesis."

In the same spirit, biologists declared in the nineteen fifties buoyed by the discovery of the double helix structure of the cellular molecule of heredity, the DNA, no doubt, Biology is chemistry, is physics. There is no need for the hypothesis of a separate nonmaterial concept of vital energy to account for life. And we will prove it by producing life in the laboratory.

Computers had just entered technology in the nineteen fifties. In the same spirit as the molecular biologists (life is the play of molecules), computer scientists said, thinking is computing. And we will prove it by producing a thinking computer who can converse with you so like another human that you would not be able to tell the difference. This is called the Turing test. Today's computers can pass Turing test.

What about feeling? We do feel, don't we? By eliminating vital energy, biologists are eliminating feeling as well since feeling is the feeling of vital energy. If there is no vital energy, where do human feelings come from? Computer scientists would rather not have feeling in science because their toy—computers—cannot feel, feeling is not computable, this much they concede. However, some bio-philosophers have theorized that perhaps feeling is an offshoot of Darwinian evolution. Maybe feeling has survival value.

Fast forward. At the time of this writing some six decades later, where do we stand with all this what the philosopher Karl Popper termed, "promissory materialism?" To repeat, no biologist has been able to produce life in the laboratory, not even programmed molecules such as the RNA without the interference of human intelligence. So molecular biology is based on a failed assumption. To top it, biologists' other pet theory, Darwin's theory of evolution of life, began to be questioned in the nineteen seventies by new data—fossil gaps. As mentioned earlier, new puzzling fossil data even surfaced very recently and philosophers and scientists had another battle of words on the Internet without any agreement.

Vital Energy

The case against vital energy is murky. Faced with the challenge from neuroscientists, biologists had to face up to acknowledging the veracity of emotions—that we have previously defined as feelings mixed up with thoughts. Biologists continue to deny the experience of feeling; they try to bypass it by defining negative emotions as "brain's response to certain stressful stimuli." Fortunately, there is no consensus on this as of yet.

Partly the problem is that in this era of the rational mind, most people so completely identify with the cortical self that they hardly ever experience feelings in the body. Some women still talk about feelings in the heart and that's about it.

I know. I was one of those thinking dominated people. What changed my mind was the following experience.

In 1981, I got a call from my fellow consciousness researchers in the psych department of the U of O. They needed me since I was a hard scientist. For testing out the veracity of a fellow who was claiming that he can demonstrate vital energy. Of course, I went. The fellow looked like a hippie but friendly; his girlfriend was with him. Indeed, he claimed that his palms were "energized," and he challenged each one of us to put our palms in between his energized palms. "You will feel tingles," he said, "That is vital energy."

Oh, yeah? One by one my psych friends went and put their palms between the palms of the fellow. But they shook their heads. Nothing, no tingles. I was the last. Imagine my surprise when I felt quite pronounced tingles; I even verified it by withdrawing my palms, waiting for a pause, and trying again. The tingles were unmistakable.

When I shared my experience with my colleagues, they wouldn't believe me. Instead, I lost my credibility with some of them who started referring to me as a mystical physicist.

If you are skeptical about feelings and vital energy, here is what you should do. Rub your palms. And then make a little gap between them and feel. You will be able to feel tingles like I did. And once you get a feel for it, it is easy to feel vital energy at the chakras, especially at the heart.

Meaning and Distinction of Human and Artificial Intelligence

The case for the equivalence of artificial and human intelligence is also murky. Yes, computer scientists have produced machines that pass the Turing test. But meanwhile, the philosopher John Searle (1994) and physicist Roger Penrose (1989) have created a new challenge. Thinking in depth is about ambiguity that meaning brings. Computers' thinking is only information processing, a computer can never process meaning from scratch. It is mathematically impossible. Nobody has been able to make a meaning-processing computer from scratch. Computers process symbols; you can reserve some symbols to denote meaning, call them meaning symbols. But then you need more symbols to tell you the meaning of the meaning symbols, ad infinitum. So, computers, even the current generative AI programs, cannot think like humans with meaning and understanding, this much is clear.

The recent invention of generative AI programs like ChatGPT forces us to acknowledge some difference between AI and human intelligence even for rational thinking. Humans use understanding to learn whereas AI learns through brute force of using an exhaustive data base that take account of all known meanings. This makes AI as or more capable than humans of rational thinking; but this is small victory. Humans can intuit new meaning; they can be creative and discover entirely new contexts for meaningful thinking. This generative AI can never do.

Meanwhile, on the experimental front, vital energies and mental thinking can now be measured (Kirlian photography, MRI). There are neuroscience experiments that establishes our free will to say "no" to conditioning (Libet, 1985).

The following question, though, remains crucial: Where do these objects of experiences come from—feeling, meaning, and purposeful intuition? Now that we have shown that the material experience of sensing can be explained by understanding the metaphysics behind quantum physics, can the other experiences—feeling, thinking, and intuition—be explained with the same metaphysics?

Bridging Science and Mysticism: Experimental Metaphysics

The materialist scientists do make a good point though. Philosophy unchecked does lead us astray. Philosophical debates about vital energy were restraining biology to open its wings to explore their science with the new technologies that were becoming available in the forties and fifties. Even today, how much energy is spent with the idle philosophical debate about if God exists. Or is there free will? Or What is life?

I referred to the two-year long debate on the Internet about What is life? if you don't acknowledge the progress in understanding quantum physics or do not bother to acknowledge the great insights of the great mystics, the result indeed are unending philosophical debates.

The point that scientific materialists miss is that not all philosophers philosophize via empty rationalization. Yes, what some people do quite legitimately can be labeled as mind f-king. Yes, ever since Descartes this has been the situation with philosophy more or less with many philosophers or it seems that way. But it is not really that way.

The problem of the human being is this: a human being even today is an unfinished project, a project in the making. It is very easy to see in the neuroscience data. On one hand, we have the clear evidence that an ordinary human being's brain alternates between a task-related area and the self-agency area; but the self here is a behavioral self, the ego, quite determined except the ability to say "no" to determined behavior; one can easily see no need to ascribe free will or creativity to it. At the same time, neuroscience experiments are revealing people whose brain often acts globally, in synchrony all over the different brain organs, with a special signature to boot. Ordinary people waking brains vibrate in the beta mode of frequency between 13.5 Hz to 35 Hz; these people's brain vibrate in the gamma mode (> 35 Hz). Read Goleman and Davidson's book *Altered Traits*.

What produced the change? That, too, is well documented. A transformational process that all the people with the different brain trait of more synchrony underwent, a process which is traditionally labeled as meditation but in truth has much more finesse in it, has creativity in it.

And here is the other point the materialist scientists miss. Popular Christianity has popularized the idea of God as a white bearded guy seating in a throne in outer space from older ideas that today is labeled as mysticism. But these mystics, according to historical record, did meditate (I bet in the same style of meditators today whose brain functions differently!), and they formulated their mystical wisdom philosophy not based on the rational mind alone but based on their intuitive and creative experiences arising in the course of their meditation. Unfortunately, they labeled their experience as revealed knowledge which in pop religion was translated as God-speak as depicted for example in the Hollywood movie *Ten Commandments*.

The truth is that these mystics were not of the ilk of mystery-loving ooh-ooh new age-y people. Entire civilizations of the past, much better than today's (by all standards except material wealth and technology), have been based on these people's teachings. I am talking about Yajnabalkya (one of the savants of the Upanishads of the Hindus), Sri Krishna, Socrates and Plato, Buddha, Lao Tzu, Jesus, Mohammad, Ibn Arabi, the Zen master Rinzai, Rabbi Hillel, etc. Contrary to popular scientists' perception, these "transformed" people described reality as a state of unity that they identified as the source of their "oneness of everything" experience. Not surprisingly, if you ask the modern transformed meditators, they will describe reality in a similar way because they too experience it the same way.

So, the mystics were not philosophers of the ilk whom the scientists justifiably abhor—the philosophers of the rational mind. What was Descartes' (Descartes is the founder in modern West of rational philosophy) real error was to base philosophy on his ordinary experience. Although his rational powers of analysis were great, he was no meditator in the spiritual sense (ironical though, his ideas of consciousness are written in a book called *Meditations* though he meant mental rational deliberations).

We must distinguish between the two brands of philosophers: the purely rational (Cartesian tradition) and the experiential/intuitive-creative followed by rational (mystical/spiritual tradition). Philosophical thinkers who were followers of the mystics (Aristotle is a prime example) created a lot of confusion because instead of taking the transformative path they took the easy way of trying to rationalize the mystical experiences of their masters and mentors with their low-level rational mind. Additionally, they wanted to popularize the teachings in their own search for power. The result is the hodge-podge of often dogmatic ideas that we call religion.

I always say, let's not throw out the baby with the bathwater. Today's scientists also perpetuate Descartes' error; they base their models of conscious awareness based on their ordinary experience. Our ordinary experience is muddled like bathwater; we can hardly see the baby—the unity consciousness in it.

So how do we resolve these issues or is it hopeless unless every scientist and philosopher becomes transformation-seeking meditators?

Help has come from an unexpected direction: quantum physics itself. The theories and experiments of quantum physics (and a few other developments in the experimental arena) are suggesting that there is no more any need to engage in rational philosophy in the old tradition. We now can make philosophy scientific. The phrase experimental metaphysics was coined by the physicist Abner Shimony. I am adapting it. Quantum physics has enabled us to verify the idea of Oneness of consciousness (the lack of distinction between you and an "other") as the basis for reality. Quantum measurement theory has given us the brain mechanism via which consciousness identifies with the brain and the brain acquires the subject/self. This idea, too, has been verified.

As previously pointed out, the big weakness of the experience-based philosophy of the wisdom traditions is their inability to explain how the Oneness splits into two and then many. Realize that that question has been answered in quantum science. This is a tremendous breakthrough.

The new quantum science also gives a theory of the creative process of transformation. The neuroscience (and a few other amazing biological data as well) verifies these details of the quantum science within the primacy of consciousness.

The best part of all this development is that this new paradigm of science makes it clear from the outset the following: The materialists are not totally wrong; they are talking about the ordinary human consciousness which *is* almost all mechanical (the only real freedom left is the freedom to say "no" to conditioned behavior). Quantum physics enables us to see that the mystics are talking about potentiality (remember: "quantum physics is the physics of possibilities") that is available to everyone but you have to actualize it to experience it and that takes engaging the brain, actually your life itself, in the creative process of transformation. The missing link between the two approaches is a comprehensive theory of life and the brain and a theory of creative transformation. All this has been done; read my books, *Creative Evolution*, *Quantum Creativity*, and *The Quantum Brain*. And now science and spirituality both can be accepted as part of a human quantum science of everything that is living and that is human.

I discovered the quantum theory of consciousness in 1985. If material objects of our experience of sensing begin as possibilities of consciousness to choose from, I soon (actually, nine years later but time passed fast in those days) realized that all our experiences must begin the same way, as possibilities of consciousness. Feelings are possibilities of vital energy movements actualized as experience;

thinking comes from possibilities of meaning. And intuitions come from the possibilities that Plato called archetypes and spiritual traditions called virtues—love, beauty, truth, and all that.

From amoeba to human, it is boundless consciousness trying to find expression within the finite. In this book I want you to get a feel for the whole struggle that life does with evolution, the society does with civilization, and the human self does through myriad episodes of reincarnation (see chapter 3) to ascend humanity to greater and greater heights.

The one idea to take with you:

> *The theories of quantum science, nay, all scientific valid theories, are discovered by creativity, going beyond rational thinking. Scientific theories, even when nonmathematical, are no mere philosophy; they literally are experimental philosophy. Their validity can be and must be verified by experimental data.*

The Quantum Self

About the subject pole of experience. The computer scientist Doug Hofstadter suggested in a book named *Goedel, Escher, Bach: the Eternal Golden Braid,* that a tangled hierarchy does appear to endow a system with a self, although for that to happen there has to be an inviolate level where there is causal power.

As an example, consider the liar's sentence: *I am a liar.* If you enter the system, you get trapped in the circularity, If I am a liar then I am telling the truth; if I am telling the truth, then I am a liar, on and on. From outside the system, you can see it is you who is creating the trap via the rules of the English grammar. So it is with our unconscious oneness. The unconscious brain has a perception apparatus to create perception and a memory apparatus to make memory, both in potentiality. In attempting to see through the brain, consciousness collapses the potentialities. From the collapsed perspective of linear time created by the memory apparatus, the two systems appear to make a trap of circularity: it takes perception to make memory, but there is no perception without memory, without the creation of one-way linear time. So, as soon as consciousness identifies with the brain, the illusion of separateness is created which it experiences as a separate self (quantum self) separate from the object it is looking at. This is how tangled hierarchical collapse creates subject-object distinction.

The real difficult paradox of quantum measurement is not who gets to choose? but this: it takes the observer to choose, but without the observer there is no actualization, no observed. This paradox is now gone too: *the observer is the observed—they come from the same unity; the observer and the observed arise together.*

There is one more paradox in how we perceive. The brain makes an electrical image of the object to be perceived, not anything that we can cognize; consciousness uses mind to give meaning to this image to have cognition just as we use our mind to make sense of electrical images on a TV screen. The idea of psychophysical parallelism is crucial here.

For all the above arguments to hold true, the perception and memory apparatuses in the brain must be quantum in some way. Ordinarily, these apparatuses of the brain would make decohered quantum possibilities. In the event of an experience which includes cognition, however, consciousness uses psychophysical parallelism in the manner of fig. 1 (see chapter 1). The individual cells of every brain apparatus or organ have epigenetic (remember: epigenetic means outside of genes) gene activators (they could be regulatory proteins) that activate the right genes to make the proteins relevant to that organ's function. The gene activation process is quantum. In this way for ordinary organ function there is quantum movement only at the individual cell level.

For the brain organs of memory and perception involved in a macrolevel experience, the individual cellular quantum gene activators of each organ involved all line up in a quantum coherent movement via nonlocality of consciousness activated via attention. This quantum nonlocality is the explanation of the so-called *binding problem*—how the different brain organs involved in an experience bind together to give the observed unity of experience.

The Ego

There is also the ego—our ordinary experience of the self, individual and personal. The quantum self, the representation of unity consciousness in the brain. loses its universality via reflection in the mirror of memory before we experience it. The reflection occurs in two stages:

1. in the first stage, a pattern of responses, learning takes place giving us habit patterns and character traits;

2. in the second stage, we behave differently in different situations in response to the same stimulus not always consistent with our character; we also reconstruct our memories a bit here and a bit there and develop various personality programs for ourselves to respond selectively depending on the situation. We have one persona when we relate to a parent, another when we are with a lover, and still another when serving an employer. In this way we become the head honcho, the boss of our programs, a simple hierarchy. This is the everyday self—call it ego/character/persona—that we experience.

The idea to take with you:

Consciousness manifests in our brain in two modalities. The quantum-self is tangled hierarchical whereas the ego is simple hierarchical; ponder what it means to you.

Hardware and Software of Life

All the above is for the human brain. Does conscious awareness have to wait till the human being comes into the picture? This would be wonderful theoretical support for the creationists' theory based on the Biblical Genesis. Unfortunately for the Old Testament aficionado, the fossil data rules this out. See chapter 4.

Let's generalize the brain solution of how subject-object distinction arises to the living cell to provide an explanation of how to distinguish between life and nonlife.

Life began with a single living cell. And then life evolved all the way to the human being. Let's assert that even a single cell has a tangled hierarchy of perception and memory built in it; it has to, in order to have the property that distinguishes life from nonlife—the ability to discern between self and environment. See later for details.

Subject-object experience begins with the first single prokaryote cell that began life about 3.8 billion years ago. That single cell evolved into organisms of multiple cells, eventually differentiated into organs in complex organisms that came later. In all this evolution and development, the experience consisted of only sensing and feeling. The meaning-giving mind had to wait till the neocortex evolved in the brain.

Why do we need the subtle worlds of possibility and the subtle experiences? This question now can be answered. The gross world of matter, macro matter, becomes approximately Newtonian. Matter makes representation of consciousness in the form of the living cell and the brain; that is what the tangled hierarchical trap is all about. A representation better be stable, not a forever moving wave like a quantum object. So, by design of the quantum laws, quantum movement is reduced to a trickle when it comes to macro matter.

But then consciousness cannot connect to matter although strictly speaking, matter still consists of possibilities of consciousness; what this means is this: macro matter cannot offer it macroscopically distinguishable quantum possibilities to choose from. The technical word for this is decoherence. In a macro-material system, technically speaking, the different parts have individual quantum movement, but the movements are not phase coherent with one another.

For the microbodies of complex living organs, decoherence stops at the level of single cells that remain quantum and functioning as an integral unit. Now the mechanism of cell differentiation comes into play differentiating the cells of individual organs according to the physiological function the organ performs (see below for details). The intermediary of the subtle—the vital and the mind (for the brain)—is needed to bring the ideas of consciousness such as survival into manifestation as organ physiology (see fig. 1).

In this way, matter makes representation of not only consciousness but also of the vital and the mental. We must think of the material part of an organ plus its physiology as hardware which obviously needs software for living experience. The vital and mental provides the software.

What about intuition and the supramental-supravital world from which they arise? For all we know, intuitions may have been a very recently arrived at experience. We know they had gods in the hunter-gatherer era of humanity; but they were all nature-gods. To associate the concept of these gods with a virtue or archetype came much later. And a full-fledged understanding of intuition is perhaps only as old as the world's first consciousness scientists when the following story appeared in one of the Upanishads.

A curious boy asks his father, a wise teacher, "Father, what is the nature of reality?" The father, though pleased with the question does not answer directly, "Why don't you meditate and find out for yourself?" The son meditates for a while, gets an idea, and goes to his father for verification. "Reality is matter—the stuff of which my body is made, the stuff of the food that I eat." The father approves. "Yes," he says, "but meditate some more." The boy goes away, meditates, and after a while, based on his experience no doubt, has another idea. "Reality is the vital body, the container of energy that I feel—vitality," he declares to his father this time.

The father approves but encourages him to meditate some more. The boy does what he is told; soon he has another idea. "Reality is mind, the vehicle with which we think and explore meaning, father." Father says, "Yes, but go deeper." The son is perseverant. This time he meditates and meditates and finally discovers intuition with shiver in his spine and all that and runs to his father. "Father, father. I found it, I found it. Reality is stuff from which our sciences come from, the context of what matter does, what vital energy movements are all about, even the contexts of our thinking; they are *Vi-gyana*, the context of *gyana*—knowledge; they also give us the values to live by." He is talking about the archetypes of course. A smile breaks on his father's face. But he says. "Good. But go deeper my son." No matter. The son is really motivated now, and meditates again and discovers the oneness of everything, that reality is one and only, with no boundary. His being fills with joy and a certainty comes to him. This is it. He does not go back to his father anymore, no need to confirm this time.

This is probably the earliest reference to four types of our experiences: sensing, feeling, thinking, and intuiting and four types of bodies that we have associated with each. Of these bodies, only the physical has a structure making it visible. The others are memories of living that we accrue as we live our life: we experience feelings making memories of a vital repertoire of quantum possibilities that has personal functional use for us individually, giving each individual a vital body. Similarly, thinking generates our functional body that we call our individual mind. We cannot make direct memory of the supramental; but we do make representations (memories) of our supramental experiences as well, with our mental and vital facilities. This gives us a supramental body, also called the soul in the West.

Add to this the quantum self—the doorway to Oneness. Oneness—the whole—can also work like a body that we all share.

To summarize the idea to take with:

Our four different experiences give us four different "bodies;" one of these—the material—is objective and external; the others are subjective and internal. These plus the quantum self—self of Oneness—make it five bodies of the human being. This is an extremely important fallout of the quantum worldview with implications for our health and happiness as well as all aspects of how we should build our societies.

The Inner Outer Dichotomy of Experiences

Submicroscopic movement of matter is quantum; but as matter makes big conglomerates acquiring bigger and bigger masses, the quantum movement gives way to deterministic Newtonian movement. The wave of potentiality of the center-of-mass of macro-objects of matter still spread when left alone, but they spread sluggishly to the extreme. So sluggishly that when you and I look at a chair, there is such little range of potentialities of position to choose from, that you and I end up choosing virtually the same position. The leeway of choice we have in this case can be calculated to be of the order of 10-17 cm, too tiny to discern except for laser-assisted measurements. Naturally, we think we are looking at the same chair at the same position; we reach a consensus that the chair is outside of us. This consensus agreement about our experience is what we call objectivity. In this way, objectivity in quantum physics is called weak objectivity; our cognition of objects does depend the observer but not a particular observer.

The case is very different for objects of feeling and thinking which are always quantum. You and I have so much potentiality to choose from that it is extremely unlikely we will choose the same vital movement of feeling or mental movement of meaning at the same time. So, since we experience feeling and thought privately, we assume that they must be internal.

In this way, the fact that we have an internal-external dichotomy between feeling and thinking on one hand and sensing on the other hand that can be explained by the quantum nature of the objects of feeling and thinking is proof positive that they are so.

The Chinese figured the quantum nature of vital energy out long ago empirically; of course, they did not call it that way. How could they, the concept was not around? The Chinese called objects of feeling *chi* and they have been studying chi for millennia. And from the get-go they noticed that feeling is experienced in two aspects: a stable or stillness aspect they call *yin*; and a dynamic aspect they call *yang*. You can of think of yin as the possibility wave mode of chi and yang as the actualized particle mode.

This yin-yang dichotomy is no superfluous thinking. It led to the development of a very successful medicine system called Traditional Chinese Medicine (TCM). Today, who hasn't undergone at least one bout of acupuncture treatment to relieve pain? Acupuncture is part of TCM.

The important thing is that because of their quantum nature, vital energies can be correlated. You can experience another person's feeling through this correlation which gives oneness and nonlocal communication between the two of you. There is much evidence that plants can pick up your feelings (Grad, 1964, 1965) as can animals like your pet dog obviously. How are they doing it? Nonlocally, of course, via the mediation of consciousness.

The Quantum Science of Feeling

I find the order in which the boy in the Upanishadic story discovered our various experiences very interesting. He discovered *prana*—vital energy—before the mind—the vehicle of thinking. Today's people would have done the opposite—discover thinking before feeling.

It is important to recognize that the Easterners had researched the subject of vital energies and feelings a lot more since the days of the Upanishad. Their findings can be called chakra psychology according to which there are seven centers in the body roughly along the spine where we mainly feel these vital energies; these are called the chakras. Researchers have also noted that all the chakras are at or near the main organs of the body, organs that do the most important functions.

It follows that vital energy has to do with organ functioning somehow. How? I got the idea of how from Rupert Sheldrake's book *A New Science of Life*, published in 1981 which proposed the idea of "morphogenetic fields" that are responsible for cell differentiation.

This was a big open problem of biology—cell differentiation. Cells in different organs behave differently; they are differentiated. It is this problem that Sheldrake was trying to solve when he creatively intuited the idea of nonphysical morphogenetic fields.

The problem is this. You know we are all born as a one-celled embryo called zygote. And then cell division makes us the wonderful body with so many organs with different functions. But every cell comes from the same original zygote and has the same DNA and the same genes and the same capacity of making all the proteins that the body can ever use.

But of course, the organs of the body *are* highly differentiated; every organ, depending on where it is, somehow activates only those genes that are needed to make the appropriate proteins appropriate for the functioning of that organ. For example, a liver is somehow instructed to activate only those genes that are needed for liver-functioning. Sheldrake is boldly suggesting that these instructions are coming from morphogenetic fields outside of the genes; morphogenetic fields are obviously epigenetic.

The idea of epigenetics is not entirely new, it came from the biologist Conrad Waddington. He conceived of the idea of an epigenetic landscape of hills and valleys for the organism's development; as the organism evolved the lie of the land would change and exert influence on the genes' instructions.

Even mainstream biologists today accept the concept of epigenetics and look for mechanisms for what genes are activated and what is left out in an organism's organ. Indeed, it is now agreed that there are molecular compounds and proteins that attach to the DNA and activates the genes, that is turn them on or off.

What directs the epigenetics? It is agreed that our lifestyle factors—exercise, diet, habit patterns and character traits—crucially affect epigenetics. But how? The clue is: lifestyle is a conscious choice.

Most importantly, cell differentiation smacks of nonlocality: how does the cell know where it is in the body or to which organ it belongs. Organ functions are purposive; how does purpose enter the picture?

To answer these questions, following Sheldrake's lead, I theorize that there are special nonmaterial fields for the differentiation of any organ for its special functioning. The function is programed by consciousness using these fields as blueprints into epigenetic quantum gene activators that attach to the DNA in the cell; in this way, only suitable genes are activated in the organ for making the appropriate proteins for an its functioning.

Sheldrake called these fields morphogenetic fields; *morpho* comes from a Greek word meaning form; genesis means creation. However, it is both the organ function and the form which are at issue, not the form alone. Therefore, a more appropriate name for these structural/functional blueprints for consciousness to use for organ differentiation is morphogenetic/liturgical fields abbreviated as morpholiturgical; liturgy is the Greek word for function.

An example. The quantum part of the physical liver (P-liver)—all the quantum gene activators—and its correlated epigenetic morpholiturgical fields as conditioned via repeated use, call it V-liver, constitute the physical-vital pair that makes liver a functional liver. Think of the V-liver as correlated vital software.

When the liver functions well, or its function stops momentarily, the quantum morpholiturgical software of the liver, V-liver, will manifest or go out of manifestation respectively. In either case it is a movement. It is these movements that we feel as feelings.

Our lifestyle, the choices we make to live our life, determines the V-organs and the feelings we have in our body as well as the actions of the epigenetic gene activators that mark the physiology of the P-organs.

This concept of the morpholiturgical field enables us to explain the chakras, a big breakthrough in our understanding of the human being. See chapter 6 for details.

The big question is: who feels these feelings at the chakras? For feelings to manifest we need a tangled hierarchy. The cortex has it. There is new data now which enables us to theorize there are tangled hierarchies and selves in other parts of the body and the brain as well, at the navel chakra, at the heart chakra, and at the midbrain chakra to be specific, the last chakra is one that we have theorized based on new data (Goswami and Onisor, 2022). Details in chapter 7.

This is breaking news for people interested in animals, specifically in this question, Do animals feel? Many of us have relationships with animals that are no less strong than many relationships with other humans, I am talking about pets. The idea that animals can feel at organs other than the neocortex (which animals don't have) is of much explanatory power for explaining recent data on animal consciousness.

Take these ideas with you:

> *Morpholiturgical fields (also called morphogenetic fields) are the organizing principles consciousness uses to generate vital software to be represented in the physiology of the biophysical hardware—the physical organs.*

> *Chakras are places along the spine where we feel the (vital) energies associated with the collapsing of the morpholiturgical fields, the source of vital software that consciousness uses for organ functioning.*

> *Do animals feel? The answer from quantum science is a definitive "yes."*

CHAPTER 3

Evolution and Purpose: The Quantum Integration of the Ideas of Lamarck, Darwin, Aurobindo, and de Chardin

Very succinctly, Darwin's theory says, Evolution is a play of chance and necessity; chance variation of a hereditary component of life (genes) and the survival necessity for the organism to adapt to large scale environmental changes that beseech the earth over the geological epochs. The last takes place via natural selection: nature selects the gene variations necessary to create new traits or improve upon the old for adaptation of the old species to environmental changes and also to evolve new species better suited for survival.

Biologists in the main cannot handle any attacks on Darwin's theory of evolution for political reasons, but in hindsight it is pretty obvious that Darwin did make a few major errors when he formulated his theory. Some of it are assumptions that did not bear out; others are errors of incompleteness as revealed by new data.

Darwin made several very optimistic assumptions. One of the assumptions, that there are hereditary molecules in the living cell, has borne out; these molecules are the genes, parts of the cellular DNA.

But the center point of Darwin's theory, of his own originality, is the idea of natural selection. Gene mutation gives variations; nature—environment with all its climate variations during the log time scale that geology gives us—provides the mechanism of natural selection for environmental adaptation. Darwin assumed nature selects from the genetic variations on the basis of survivability. Gene mutation is slow and incremental bit by bit. Hence according to Darwin's theory, evolution should be slow, incremental, and continuous.

Natural selection and survival are enough for explaining how a particular species adapts to environmental changes in the long-term geological scale. Darwin *optimistically* assumed that there is also enough mechanism here to explain speciation—evolution of one species to another. Indeed, microevolution—evolution with only little changes involved is found to be consistent with Darwin's slow and continuous evolution.

Thus, initially, the fossil data was assumed to agree with Darwin's theory. But of course, there was new more detailed data. First the fossil gaps were revealed; there are major discontinuities in our fossil record. Additionally, the fossil records—their time sequences—are now in some turmoil in view of new data.

In this way, the new data calls forth major revision of Darwin's theory and Neo-Darwinism.

The third assumption Darwin made is that there is no purpose in evolution. He himself did not like this assumption, but he was stuck. Newtonian science is cause-based; purpose has no role to play. Darwin himself was very aware that an organ like eye, in addition to survival value, has purpose—it gives living being a broader perspective of the environment eventually bestowing us humans the capacity of forming a worldview. Darwin was also aware that a gene mutation leading to perhaps the development of one thousandth of an eye has no survival value; it would not pass natural selection. How would these small increments accumulate then? they couldn't. And yet he was prompted following Newton's causal laws of material movement to theorize that the eye evolves from little increments (one thousandths of what is needed at a time) because he wanted to use survivability as the only causal driver of evolution rather than a purpose giving additional impetus and mechanism.

As mentioned earlier, there is fossil data showing that there is an arrow of time in biological fossil record which goes from simple to complex. Could this be evolution's purpose, to produce complexity? Complexity allows better ability to process not only survival needs but higher needs of living.

This is the point. Survivability is a tricky concept. Bacteria that evolved early have enormous survivability, if you are thinking of quantity, like longevity. We humans with higher needs do not live very long, but what an enormous difference in the quality of living we have compared to bacteria.

I have already argued that conceptually, Darwin's theory and molecular biology are in a collision course, and it is molecular biology that is wrong; survival is not a molecular property, biology is not chemistry. Indeed, we will see (in chapter 6) that Darwin's mechanism of survival necessity, when looked upon from a primacy of consciousness point of view, has some limited but important contribution to make to macroevolution. On the whole, Darwin's original theory has to be modified to be sure, but in the same sense that Newton's theory is to be replaced by quantum theory. In a certain limit, for macroobjects, Newtonian physics retain validity; similarly, in some limiting situations—microevolution—Darwin's theory retains validity.

The most glaring shortcoming of Darwin's theory is that variation takes place at the micro genotype level, but natural selection must take place at the macro phenotype or organ level. Moreover, developmental features in early evolution keep repeating in later stages suggesting that evo-devo, evolution and development together is the real story of evolution.

As mentioned before, Lamarck's theory of evolution which proposes evolution via development was rejected because biologists found no mechanism for hereditary continuation of developmental changes. The central dogma proposed by the biologist Francis

Creek more or less stands: information can only flow from genes to proteins but not the other way from proteins to genes. How does developmental information propagate then from generation to generation of living species? We now have discovered the mechanism.

A theory of evolution based on quantum biology and the primacy of consciousness solves all the problems of Darwinism. It gives a quantum mechanism for explaining the fossil gaps of macroevolution; and yet, for microevolution it smoothly corresponds with Darwin's slow and continuous evolution. It provides a mechanism for the propagation of developmental changes over time and thus explains evo-devo.

Most importantly, the new approach connects Darwin's and Lamarck's ideas with those of Teilhard de Chardin and Sri Aurobindo. If you have an open mind, the new integrative theory of quantum evo-devo really resolves one of the fundamental sources of controversy between science and Christianity (evolutionism vs. creationism—how God simultaneously created all creatures and maintained them through the great flood) solving at once the problem with the politics of evolution, a major source of today's worldview polarization.

The new theory, having incorporated purpose and the directionality of time, can also tell us something about which Darwinism is necessarily mute. The new theory can tell us the about the future of evolution.

Some Details of Aurobindo and Teilhard de Chardin's Evolutionary Ideas and How Quantum Science Scientizes these Ideas

The idea that consciousness (the causal level of reality) comes to matter for representation via the intermediary of the subtle is very old, going back to the Vedic times in India. The mystic/philosopher Sri Aurobindo developed this idea into a consciousness-based theory of evolution. Aurobindo suggested if consciousness remembers everything, it is impossible for the unity to split itself into two—subject and object—even in an illusory way. So, he proposed that consciousness has to forget first, has to progressively impose limits on its omniscience, a process he called involution. The final stage is matter, complete forgetfulness; and now, according to Aurobindo, evolution begins consisting of progressive remembering.

There are also even more naïve models where it is proposed that consciousness progressively condenses from causal to subtle to gross.

The ultimate state of reality (the causal level) is eternal and unchangeable, that makes sense, right? Otherwise, we can always ask, what was before? Even the early Vedic rishis were puzzled you know. There is the following verse:

Who knows for certain? Who shall here declare it?
Whence was it born, whence came creation?
The gods are later than this world's formation
Who then can know the origins of the world?

None knows when creation arose
And whether He has or has not made it.
He who surveys it from the lofty skies,
Only He knows—perhaps he knows not.

In quantum science, the ultimate origin—the causal domain—is a state beyond science, beyond qualifications and is beyond its scope. Then cause unknown, involution happens first with the monistic idealist ontology: nondual consciousness and its possibilities, this begins the subtle stage. The possibilities progressively involute, becoming less and less subjective, first the archetypes, then meaning, then the vital. And finally, the completely objective possibilities of matter come about. This finishes the subtle stage of involution. Note however that there is no real time in the domain of potentiality; the convolution does not take place progressively in real time.

In this way, the quantum picture introduces the subtle first in terms of possibilities and then manifestation via collapse, and evolution. This makes clear what Aurobindo was trying to say.

And evolution occurs in the reverse order. First matter evolves still in potentiality until the living cell with tangled hierarchy comes about with the embodiment of consciousness and manifestation—gross world of space time experience begins as the ideas of consciousness are embodied in matter via psychophysical parallelism. Life and the ways the living functions and feels evolve; second, the ways we impose meaning, that is the mind, evolves, and last, the archetypes themselves are manifest in us through the intermediary of meaning and feeling. (As mentioned before, our brain can only make memory representations of the meaning and feeling, no direct representations of the archetypes.)

So, there is a future of evolution in this theory. The Anthropological studies do indicate that we have gone through the early stages of evolution of life and are now getting ready for archetypal embodiment (see chapter 8).

Purposive Life

What is the purpose of all this? Now this often-asked question can be addressed. Consciousness evolves in order to know about what its possibilities are and how well they can be manifest within a set of scientific laws created for this particular purpose. Life is a learning journey. The feelings give it the passion to live; meaning of life comes from the mind; and life becomes purposeful with higher needs as we explore and embody the archetypes with the help of the vital and mental. We call this latter stage soul-making.

Consciousness brings purpose into the laws of manifestation through evolution. Darwin struggled with the question: what is the purpose of a single gene mutation good only for 1000th of an eye? The quantum approach says: none. So, one thousandth of an eye never manifests. Instead, it is allowed to accumulate in the unconscious as possibilities and processed there. All new manifestation waits in potentiality until a purpose can be served, such as the making of a new organ with a new function.

How do the forms thus developed at the macrolevel become hereditary? Via the phenomenon of nonlocal memory connected with the phenomenon of reincarnation.

Reincarnation is now empirically established feature of human child development. The theoretical problem of what reincarnates has also been solved (Goswami, 2001). The memory of learning—the propensities we grow as our habit patterns and character traits—is nonlocal: it resides not in the brain but outside of space and time altogether. Below we develop the details.

Reincarnation

The question, "Is there a surviving entity after death?" is very much alive even after decades of rampant materialism in our culture. The question implies the assertion, death is only when the surviving entity leaves the body, which is the religious position on this matter. This is why doctors and hospital administrations are ambiguous about whether or when to terminate the life support systems of brain-dead patients under prolonged coma. In view of recent data on near-death experiences, such caution is justified to a degree.

Religious traditions usually use the ambiguous word 'soul' for the surviving entity. To avoid ambiguity, in quantum science we use the phrase 'quantum monad' for it.

When we base our science on the primacy of consciousness which quantum physics demands that we do, then the question of survival takes a new turn. With death the material body, the brain is the first to go and then the rest, the organs do not function anymore, and the entropy law takes over leading to decomposition. However, consciousness and its other personalized lived possibilities—vital, mental personal (and universal, but that is of no concern here) software—remain intact having decoupled themselves from the body and the brain; there is no entropy law in these domains.

The intriguing question then becomes "Is there any "me?" in what survives the death of the physical body. Let's examine this question in some detail.

One more comment in passing. Western religions use fear of hell as a deterrent for people going astray from morality; in the East, they use fear of reincarnation as lower animals. That reminds me of a story.

A fellow gets very drunk and is barely able to return home, get undressed, and lie down by his wife on their bed and in no time falls asleep. Soon he has a vision. He is at the door of heaven with St. Peter at the gate. St. Peter look at his ledger and says sternly, "Nope, you have committed too many sins. No heaven for you." The specter of hellfire and brimstone flashes before Peter's eyes and he pleads for mercy. St. Peters relents a little, "Ok, you can reincarnate, but only as a chicken. Do you agree?"

They did have a chicken farmer near their house. So, he will at least be able to see his beautiful wife occasionally, thought Peter. He said, "OK." In a flash, he finds himself as a chicken talking to a rooster who says, "You are new, aren't you?" "I guess so, says Peter. "Hey, my stomach feels kind of funny!" The rooster laughs. "You are about to lay an egg. Go ahead, do it. Squeeze it out." So, Peter did, and it felt great. Why not again? So, he did it again, and another egg dropped out.

Suddenly another flash, and he found his wife screaming, "Peter you drunk! You have shat in our bed again!"

What part of "Me" Transmigrates from Incarnation to Incarnation? Nonlocal Memory of Learning

In the foregoing, I referred to the problem of individual ego development and stated that it arises from conditioning while growing up. Let me elaborate some details of how that conclusion is reached.

The physicist Mark Mitchell and I (1992) studied the quantum mechanics of the repeated measurement of the position of a quantum object in confinement and found that since any quantum measurement involves memory-making, subsequent measurements involve not only the processing of the present stimulus but also of the past memory. (We pretended for reasons of mathematical feasibility that we were dealing with an electron, but of course, the real applicability was for the quantum object of the mind or the vital.) The feedback of the past memory gives rise to a nonlinearity in the otherwise linear quantum equation of movement of the electron. Don't worry of the technicality of the concept, but the nonlinearity leads to a modification of the probability of the quantum object's response to the new measurement in favor of its past responses. In this way as a result of repeated feedback from past memories the object becomes "conditioned" to show up in the same position that it showed up before. That is, it behaves more and more like a Classical Newtonian object forgetting its quantum freedom.

When this result is generalized to the quantum mind and objects of meaning, we must conclude that reflection in the mirror of our past mental memory has the same effect of producing conditioning: we respond to a previously experienced ("learned") stimulus as we had responded before. This produces a pattern of habits of mental response. Even the archetypes that set the contexts of our mental thinking, become personalized to a degree, an aspect of ourselves that we call our learned abilities, that we have learned to learn, one step beyond rote learning. This gives us our mental habit pattern and eventually, some of our habits become entrenched as our ego/character.

What happens is that as we get confident about what we have learned, we relax while practicing, and inadvertently fall into the quantum self. When we get back to our ego, we have gained conviction about the trait we have learned, it has become a character trait.

The point to note is that the habits and character traits originate as modifications in the quantum equation of movement of mental objects; this mathematics comes from the scientific law of mental movement. Where do the laws reside? The law of gravity is not software in a falling body; it is a universal law, a part of the archetype of truth. And so it is with the mental law of conditioned habits. Their memory is archetypal, nonlocal, for which the Sanskrit word is *akashic*. I mention this because the Vedantic theoreticians of the Indian theory of reincarnation described karma as *akashic memory*.

A neurophysiologist named Karl Lashley was trying to find the location of memory in the rat brain when rats learn a behavior. You remember, rats can only cognize via feelings and vital energy; however, the same conditioning dynamics as the mental takes place in the vital.

Lashley taught rats how to find cheese in a Y- maze. In one branch of the Y there is cheese, the other branch of the Y gives an electric shock. Of course, the rats very quickly learn how to find the cheese. In order to find the memory of where that learning is located, Lashley started chopping off parts of the rats' brains.

If, after chopping, the memory is gone, then the conclusion is that the memory must have been in that chopped portion of the rat's brain. But with five percent chopped off, and mind you, different rats lost different five percent portions of the brain, nothing happened. Basically, all the rats could still find the cheese. So then, Lashley started chopping off ten percent. Twenty percent. Even with fifty percent of the brain chopped off, the rats could no longer see, could no longer walk, but they crawled blindly and found the cheese. Lashley concluded that the memory of learning, the propensity of finding the cheese must reside all over the brain.

Can the brain be something like a hologram having this property of being all over? Holograms take special laser mechanisms to make; such theories of the brain are not credible from an evolutionary or developmental point of view.

An equally valid conclusion would be that the memory is not in the brain at all. It is *akashic* nonlocal memory confirming what is suggested above giving empirical validity to the quantum science deduction.

The mind (or the vital), as stated before, as even Descartes knew, has no micro-macro division, no structure. So, there are no individual minds or vital body due to structure. However, the conditioned habits and character traits of the response pattern that we all in our egos identify with along with our mental and vital history recorded in correlation with the brain memory gives us a functional individual mind and vital body each with large memory/learned repertoire (software) with habit patterns and character traits to use the repertoire.

Now the crucial point of survival after death can be made. When we die, the brain dies with its memory of personal history. But the nonlocal memory that is represented by the character and habit patterns remains safely locked into the appropriate nonlinear quantum equation of movement that comes into play whenever these character patterns are used via access to a material nody. This nonlocal memory is not part of the brain's physical neural correlates of mental and vital memory. It does not die with the brain. It is nonlocal quantum memory.

Obviously, quantum memory survives death of the physical body. The package of surviving quantum memory of the mental and the vital character traits/habit patterns can be recognized what is traditionally called the "soul" and we call "quantum monad."

Take these ideas with you then:

> *Learned habits and character traits of the human ego are stored outside of space and time as nonlocal memory available in future times and places;*

> *Law of karma: There are whole bunch of incarnate human beings across space and time as in a string of pearls correlated in such a manner as to inherit the accumulating*

memory of all the previous incarnations. Additionally, nonlocal access to the learned repertoire associated with the habit or trait is also allowed.

Nonlocal memory brings us a dharma—a personal archetype to explore—along with the karma—habits and character traits—that facilitates the archetypal exploration;

Following your dharma is like following your bliss.

Materialists sometimes worry that "certainty about the next life is simply incompatible with tolerance in this one." I am quoting the philosopher Sam Harris. I think there is some truth to this concern. Indeed, too much concern about the other world has made certain religions especially negligent about this world. But this concern is misplaced because the data indicates in favor of not only survival after death but also rebirth. You cannot escape this world easily!

In fact, striking and very solid data exist for the survival of a disembodied quantum monad of a person taking "rebirth" in a new physical body. The psychiatrist Ian Stevenson (1987) at the University of Virginia Medical center collected compelling data for reincarnation over many years. The data mostly consists of accounts of past-life recall of children of various countries of both East and West that were carefully followed through and verified.

This kind of data indicates, to say the least, a nonlocal correlation between a currently living human being with many others that lived in the past and will live in the future as if they are part of a chain of pearls. This is the law of propagation of karma. Using this nonlocal correlation, a future incarnate body is privy to the use of the character of the incarnate being that lived in the past. It is called the reincarnation of the past person. *This transmigration of (mental and vital) character/habit patterns from one incarnate human to another dictated by the law of propagation of karma is what gives us the quantum model of reincarnation.*

There is one difference of our model with the traditional Vedanta model. The child is not born with all the character traits and habit patterns at once; in quantum science, they are potentialities waiting as available for the child from the moment of the new birth; they are embodied only when there is a need. The need acts as a trigger.

The quantum model can be verified by examining if character traits and habits are triggered in children in the period of child development effortlessly (sometimes even during adult development) that cannot be explained as due to either nature (genes) or nurture (sociocultural influence).

Indeed, Stevenson's data have ample examples of the inheritance of such acquired characteristics through reincarnation, data on geniuses, data on autistic children suddenly becoming "savant", data on people with unexplained phobias, the data on people with unusual language ability, etc. Read my book *Physics of the Soul* for details of such data from Stevenson as well as other researchers.

Reincarnation is often a conceptual part of Eastern religions, but it is not restricted to them alone. Reincarnation was very much a part of Christianity until the fifth century A.D. when the idea was eliminated from official Christian teachings for political reasons. And as the philosopher Geddes McGregor (1978) has cogently argued, Christianity indirectly retained the idea of reincarnation in the notion of the purgatory after death, a place where the surviving soul waits and learns. Learns how? Via reincarnations.

Among our propensities that we learn, the most important one is creativity. The capacity to create (a quality called *guna* in Sanskrit) increases both quantitatively and qualitatively in our learning journey through various reincarnations.

There is evidence also for something else. Every time we reincarnate, we have a favored archetype to explore; it is called *dharma*, spelled with lower case d. The secret of motivation toward transformation lies in finding your dharma, exploring which gives you bliss. And then life is simple. As the mythologist Joseph Campbell used to say, Follow your bliss (dharma).

Back to Evolution and Purpose

Darwin pondered a lot about how to include purpose in the story of evolution, how to integrate Lamarckism in his theory. To no avail. Even with the conceptualization of evo-devo, that evolution and development take place hand in hand, biologists have found it difficult to find a hereditary mechanism for learned propensities to propagate in time. Quantum science is suggesting that mechanism—reincarnation.

The point is this. As pointed out later, evolution displays fossil gaps for which the quantum explanation is discontinuous quantum leaps precipitating all that accumulated gene mutations as they become ready to make a complete organ of purpose. This happens quickly leaving no time for fossils to build up; hence the fossil gaps. Now realize that these quantum leaps of creativity require a lot of a protracted process for precipitating it. The creative process has four stages: preparation, unconscious processing, sudden insight of quantum leap, and manifestation. Of the four stages, the first two—preparation (doing such as imagination) and unconscious processing (non-doing or relaxing)—are to be practiced in tandem that I call do-be-do-be-do (Goswami, 2013).

It is a fact that single animals, even dogs, cannot handle ambiguity inherent in the creative process. Animals do it because in these creative situations, their consciousness become group consciousness. They handle the ambiguity and take quantum leaps as a group, as a connected nonlocal unity. This is why the memory of quantum learning in evo-devo is nonlocal group memory—unconscious memory in potentiality that the entire group and eventually the entire species can access—something akin to our collective unconscious; developmental propensities are inherited through manifestation of nonlocal memory by future generations of the species from their collective unconscious.

Only in the course of human evolution, when we develop the rational mind and lose our ability to nonlocally connect as a group long enough to engage the creative learning process, reincarnation becomes an individual affair.

So, realize this and take with you:

> The laws of consciousness and evolution include reincarnation. Death is not an end but a renewal to try again.

> Quantum physics in the form of the concept of nonlocal memory gives us the mechanism for the continuation of our learned propensities from one life to the next.

> It is the quantum nonlocal memory of group consciousness that propagates phenotype form and function from generation to generation.

PART 2

Purposive Life and Its Creative Evolution: Evo-devo

CHAPTER 4

Molecular Biology's Errors and How to Correct Them

Molecular biology is one of the important dogmas of scientific materialism—Biology is chemistry. At the time this assumption made its way into science, it was a bold assumption and in science sometimes this kind of assumptions pays off. But in this case after more than sixty years of trying it out, its errors and failures should be acknowledged, and the assumption should give way to new ideas so biology can progress further. Instead, the tendency among biologists has been to hide its errors.

It is true that molecular biology has helped the development of biotechnology and bioengineering with some success. But if there are basic errors in the theory, can engineering based on it be fully reliable? There is enormous controversy with bio engineered food and vaccines for example.

What are the two biggest errors that has made biology into a risky engineering science? The first one is its very fundamental defining assumption that there is no difference between the living and the nonliving, that interaction among elementary particles of matter can produce the complex assembly that can reproduce and maintain itself and can even evolve, that we call a living cell. Why is this assumption wrong, an error? Because more and more evidence is accumulating that a living cell can cognize: it has subject-object split experiences; it can distinguish between itself and its environment and between likes and dislikes. Nonliving objects cannot cognize.

In other words, biology of living objects cannot be reduced to chemistry of nonliving objects. There is a joke on the Internet. Why do physicists divorce biologists? Because physicists have no chemistry with biologists. Alas! It is not a perfect joke. The correct statement is: Newtonian physicists have no chemistry with biologists.

Quantum physics brings physics' connection with biology back via the concept of tangled hierarchy.

Initially, when I discovered that the concept of tangled hierarchy in the brain can solve not only the quantum measurement problem and the problem of how the brain acquires a self and acts like the subject pole of experiences but also the paradox of how the brain cognizes, my initial thinking was that the tangled hierarchy is unique to the brain. But this thinking has one interesting consequence. There is no collapse of quantum possibilities before the brain possibilities came about in the unconscious so that the human observer can manifest.

Let me elaborate. Nonscientists always find it a paradox that a conscious observer with a brain is required to collapse the possibility waves. They simply ask: there was no conscious observer when the big bang happened; it was so hot then. Right, there wasn't. However, in quantum physics, the problem is solved by the concept of delayed choice: as the quantum possibilities of the universe evolves in possibility and eventually the brain comes together as possibility and collapse can happen, all that went before, all the potentiality that was causally necessary to make that moment possible, collapses too, but retroactively, going backward in time. That means, the big bang would collapse in its own time, some 13.8 billion years ago, galaxies a billion years later, and so on. Time is a parameter, a label, for the possibilities evolving in time in the unconscious.

Realizing this, I had to face the facts of the fossil records. If the collapse first occurred after the human brain came about in potentiality, then there would only be fossil records which are causally connected to humans. There would no fossils outside of this one lineage. But you look at the fossil record and you will find that this is just not the case. There are many lineages in the fossil record that do not end up in a human being, even extinct lineages.

This could mean only one thing. Quantum collapse must have occurred first when the simplest living cell—the prokaryote—came about in possibility. Even the simplest living cell is capable of experience as it seems from other experimental data as well. The most glaring one is of course that a living cell can distinguish itself from objects of its environment. It can sense objects as separate from itself. It can even cognize whether the sense object is harmful or nourishing via feeling it.

Sympathizers of creationism behold! Who creates the world and life in it? Undoubtedly, the causal agent of quantum collapse via delayed choice in conjunction with the first manifest living cell. If you call this causal agent God and see the act of collapse as downward causation by God who is to object? However, *The book of Genesis* idea of all creation happening all at once a mere 6000 years ago has to be given up.

The second big error of molecular biology is the assumption that material forces and laws that produce hardware of the living somehow also manage to produce the software of what living entails. Never mind that software is purposeful (they smack of what Aristotle called final cause) while the hardware forces are purely causal (initial cause).

From the quantum science perspective, these errors can be corrected. If consciousness is the ground of being, we now know what is required to make hardware so that it can represent consciousness and acquire a self: a tangled hierarchy of cognition and memory apparatuses. So, the question What is life? from this perspective becomes "where is the tangled hierarchy in the living cell?"

From the quantum science perspective, the answer to the second error is also straightforward. Matter makes hardware; hardware dynamics is governed by material laws. Software is needed for performing purposive functions. Yes, purpose in addition to causal laws. Consciousness injects purpose into material hardware of physical flesh using nonmaterial fields that give the blueprints of purposive function. As previously discussed, I call these fields morpholiturgical fields; reminder: the Greek work liturgy stands for function.

Now let's look into more details.

Living Cells

Single cell creatures are of two kinds: the prokaryote (non-nucleated) and the eukaryote (nucleated).

The two most important components of the cell are the DNA and the protein. A third important ingredient is the ribonucleic acid (RNA) molecules that mediates the information exchange between the DNA and the protein. A fourth important ingredient is the cell membrane giving the cell a confined structure: an inside and an outside. The cell membrane is made of fatty acid molecules, such as lipids. A fifth important ingredient is the cytoplasm that fills the cell body and facilitates transport of the proteins and that also acts as a solvent.

Eukaryote cells are nucleated. What this means is that the DNA and the machinery that conveys its information to the RNA are sequestered by a distinct membrane; a cell within a cell so to speak.

Let's discuss the complexities of some of these cellular molecules. A protein is a large polymer chain of amino acid molecules, so complex that nobody has ever succeeded or can conceive of ever succeeding in manufacturing it in the laboratory starting from the amino acids. In the cell, of course, DNA has the code (the genetic code) to assemble the amino acid sequences to make proteins.

As such, proteins are instrumental for producing the DNA in the cell as well. *This circularity, DNA makes proteins (with the aid of the RNA), and proteins are needed to make DNA makes the origin of life problem impossible to solve for the materialist.*

But in truth, the conceptual poverty of molecular biology is exposed the best by considering the question, What is life? We can look at a living cell and a moment later, if the cell dies in that moment, at a dead cell. As far as molecules are concerned, nothing has changed, the macromolecules of life are still all there. How do you go about explaining the difference? In quantum biology, the difference is simple: a living cell is conscious; a dead cell is not; consciousness has withdrawn, this is what death is about.

Are Single Cells Conscious and Capable of Cognition?

The cognitive scientist Arthur Reber, in his book *The First Minds*, agrees that a single cell is conscious and capable of cognition. And he cites data in favor of the idea.

The biologist Herbert Jennings exposed a microbial protozoan, *Stentor roeselii,* a tiny single-celled organism, commonly found in a pond or ditch, to an irritant, carmine powder. Amazingly, the creature actively tried to avoid the irritant by shielding its body and staying away from the irritant. This illustrates cognitive ability.

Just to be sure, more recently, Harvard researchers repeated the experiment; this time they used a different irritant—microplastic beads. No surprises. They found similar avoidance behavior. There is no doubt that stentor roeselii can cognize; it has experiences of sensing and feeling; in other words, it is conscious.

What is Life? More Details

How molecular biologists deal with the question, What is life? reminds me of a Buddhist story. An inquisitive fellow creates havoc among the gods in heaven by asking the question, "Where does the universe end?" Finally, the creator himself, God, appears. When the fellow repeats his question, God takes him aside and says, "These gods look to me with such reverence that I could not say this in their presence. You see, even I do not know where the universe ends."

In my mind's eye, I visualize a fellow of modern times going to a famous university's biology department and creating havoc with the question, "What is life?" Some would say, self-maintenance that is a function of the proteins that defines life; some would say that it is self-reproduction for which the DNA is the main player defines life. But the fellow is not satisfied. A crystal can maintain its form. And a fire in the prairie tends to reproduce itself. But we wouldn't call those systems alive, would we? Neo-Darwinists come and say, "Life evolves; evolution is what separates life from non-life." But the fellow is patient to counter, "There you go. If life is just nonliving matter, it must evolve in the direction of increasing disorder or what is technically called entropy; all matter evolves this way. But life evolves against the entropy law, in the direction of increasing order. So, life must have something other than matter as well." Some more modern sophisticates conversant with physics and chemistry would say, aha, we have an answer to that. It is true life evolves and order increases; but you see life needs solar energy. The sun pays for the increasing order of life on earth by increasing its entropy. You have to apply the entropy law to the sun and the earth together which makes a closed system." The fellow is still not satisfied. "Prove to me that the thermodynamics of an open system really can go uphill against entropy and produces and evolves life." Now the expert shuts up. He knows that all such theories have bitten the dust.

So it is that in my imagination the Buddhist story repeats itself. The Head of the department herself comes, takes the fellow aside, and says, "My colleagues hold me in such esteem that I did not want to admit this in public. Also, any public admission of the shortcomings of biology would only put more ammunition in the hands of those creationists. More court cases, who needs them? But you see, we biologists still do not know how to define life, we skillfully avoid the question."

She chuckles. "We have been clever. *We get by listing a whole bunch of properties as characteristics of life hoping that no one will notice that some of these characteristics are impossible for material objects to fulfill."*

Okay, so let's list these characteristics. In one biology textbook, I noticed the following plethora of properties mentioned in order to distinguish life from nonlife:

- Organization: Living systems are organized starting with the cells, to multiple cells or tissues, to organs, to the organism.

- Metabolism: Living systems are capable of maintenance through metabolism—taking energy-rich material and eliminating, on the average lower energy material. Some of the energy acts as the fuel of living processes, some accumulates and is released after death.

- Homeostasis: Living systems can maintain homeostasis even under threatening environmental conditions by at least partly converting toxic chemicals into less harmful ones.

- Selective response: Living systems can respond to stimuli selectively. They can identify what is food and can move away from offensive stimuli.

- Growth and biosynthesis: Living systems go through phases in which they take in material from the environment and grow.

- Genetic material: Living systems contain hereditary information in the form of genes (portions of nucleic acid molecules—DNA and RNA) from previously living systems.

- Reproduction: Living systems can reproduce new living systems similar to themselves by transmitting some or all of its genetic material.

- Population structure: Living systems (organisms) form populations—groups related by common ancestry. Organisms are capable of sexual reproduction; the organisms of a population can interbreed.

This is the best a materialist biologist can do if he doesn't want to admit his ignorance or shortcomings of his theory in public. In this way, many biologists today are convinced that they must put up a front that they understand what life is according to the terms mentioned above. Life needs only molecular biology for definition, nothing more.

But truly, look at some of the items of the list, and you will find with a little introspection that it packs a punch that can knock out the metaphysics of scientific materialism altogether. In other words, the idea seems innocuous in a list like this but involves something beyond the capacity of scientific materialism to explain.

For example, take the item selective response: a living being's characteristic of being able to tell food and distinguish it from what is harmful is truly remarkable because, as argued above, it underlies the cognitive capacity of distinguishing itself from the environment which matter does not have. But this ability to distinguish is skillfully hidden by labeling differently.

This self-distinction is the central defining quality of life. This is the quality that goes away when the cell dies.

The same marketing trick is used under the item *growth and biosynthesis*. Yes, a living being does have this ability of distinguishing itself from the environment, even details like what objects of the environment will help it to grow and what objects would harm it; but how does it get this property of distinguishing itself from the environment? That question is avoided by skillful phrasing.

What distinguishes life from nonlife is really that life has a self, an innate ability to distinguish itself from its environment, other objects. Matter consists of objects; objects interact with other objects when close, no preference. How can a conglomerate of material molecules like this generate anything that can discern between itself and environment?

So How Does a Living Cell Get its Self-Identity?

From the quantum measurement theory, we learn that it is the tangled hierarchy between brain's organs for sense perception and its cognition by the mind on the one hand, and memory on the other hand, that gives the brain— a material object—the capacity to acquire a self. From what we know about a single living cell, it too has all these capacities of sense perception, cognition via feeling, and memory. Even a single cell can distinguish between itself and the environment; in other words, it can sense the environment as separate from itself. It can even cognize the nature of what it is sensing via feeling good or feeling bad. A living cell can also remember: what objects of the environment are friendly (food) and what are enemy (poison) once it encounters them a few times. So manifest consciousness or awareness—subject-object split—is present in a living cell. We can also surmise that manifest consciousness began on earth with the creation of a single living cell.

A living cell has programmed molecules—DNA and protein. DNA is the hereditary molecule. Proteins have built in programs to carry out purposive cellular functions.

The biologist Bruce Lipton (2005) has shown that even the prokaryote cell has programmability in its protein structure of its cell wall. Accordingly, consciousness programs the protein structure in the cell wall for suitable gene activation for making proteins needed for the basic biological function of perception-cognition and memory giving it the experience of rudimentary sensation using the blueprints of suitable morpholiturgical fields. The memory-making in the cell probably also uses the memory-making capacity of water in its cytoplasm.

The movement of this morphogenetic/liturgical field gives the cell feeling, ability to cognize.

I repeat. Molecules are governed by physical laws, no doubt about that. But physical laws do not have purpose built into them; they are causal laws. Who brings purpose to the cellular molecules and the cell itself? Consciousness does. And consciousness uses blueprints—morpholiturgical fields—to program biological functions in the macromolecules of the living cell and the cell membrane. The movement of these morpholiturgical fields produces the vital energy that a cell can feel. That is how a cell cognizes, with the help of the feelings of the vital energies. Proteins help it sense and the feeling helps it to know what the sensing is about.

For a tangled hierarchy, we also need the cell to have a memory mechanism. Does the cell have the capacity of making memory? Proteins help but where is the memory located? The water of the cell's cytoplasm has memory-making capacity for the movements of vital energy. The crystalline property of water changes with vital energy memories (Emoto, 2004).

In this way, a living cell has both cognition and memory apparatuses that form a tangled hierarchy which the cell uses to embody consciousness. In this way the cell acquires a self. Life explained.

Behold! Getting rid of the straight jacket of molecular biology—biology is chemistry—and replacing it with quantum biology has given us a biology that includes feeling. What do cells cognize with, give interpretation to sensing? They cognize with feelings. Indeed, until the neocortex comes into the evolutionary picture, all living composites of cells cognize with feelings.

Summary: The Distinction of Living and Nonliving in a Nutshell

So basically, in more detail what is the distinction between the living and nonliving? The difference is:

- Living beings can manifest a self, a representation of consciousness itself. Most importantly this gives the living system autonomy and the ability to distinguish between self and environment, what is food and what is harmful.

- A living being has experiences that allow it cognition—sensing and knowing.

- Living beings without a neocortical brain cognize with sensing and feeling. And since a cell always comes as physical-vital pair, sensing and feeling would always come mixed up in the cellular experience all the way up in the evolutionary ladder until we develop an additional way of memorizing in the neurons.

- Living systems, by virtue of its two macromolecules—protein and DNA—can carry out certain biological functions; the immediately obvious ones are maintenance and reproduction—survival needs. It's in the nature of the movement of consciousness, a play of conscious purpose that life evolves.

- As life evolves, cells organize into multiple celled tissue and organs of specific purposive functions. Since organ functions are purposive and not just cause-driven, they need purposive vital software for guidance of the living hardware.

- With the neurons of brain coming into the scene, self-identity at the level of the organ aside from cellular consciousness, appear in the scene, and living systems (primates, cetaceans, and humans) carry out functions that satisfy higher needs associated with more sophisticated feelings. As the brain evolves the neocortex and with it the ability of satisfying even higher needs requiring higher ability to cognize, such as thinking and intuition.

- Living systems form populations which is once again an expression of the movement of consciousness an aspect of which is nonlocality.

- Humans are the epitome of evolution so far. Why? its cognitive abilities are much more versatile that any other living creature on earth.

Concepts to keep with you:

> *Molecular biology—biology is chemistry—is categorically wrong; this assumption cannot explain life, cannot distinguish life from nonlife.*

> *Materialist science is a valid science of the nonliving universe but is only an incomplete science for the living.*

> *Quantum science does distinguish living and nonliving via the concept of tangled hierarchy.*

> *Most important: the elaborate definition of life given above.*

Molecular Biology's basic assumption is Wrong for biology and incompatible with Darwin's theory of Evolution

The question What is life? really exposes the conceptual poverty of the basic assumption of molecular biology: life is chemistry is physics. We cannot define life without having other nonmaterial organizing principles such as the morphogenetic/liturgical fields and of course consciousness itself.

Biologists can convince themselves about the wrongness of molecular biology's basic premise above in another way. Try to explain Darwin's theory of evolution with it. Molecular biology and Darwin's theory are supposed to be the two bulwarks of materialist biology; they must be compatible. Are they?

The central point of Darwin's theory is life's survivability: organisms strive to survive and the fittest to survive are selected by nature and propagate. If organisms are nothing but molecules, then this survivability should all the way be traceable to the molecules.

Origin of life, how life originates, is another big concern of biologists. Some theorists thought if Darwin's theory is brought all the way to the molecular level, maybe it will be easier to solve the origin of life problem.

The Second Error of Molecular Biology: do molecules have survivability?

If you cannot solve the problem as is, reductionists know that it is wiser to redefine the problem and solve the redefined problem. Originally, the researchers were trying to make plausible models for the origin of life as we know it—the DNA-RNA-protein conglomerate confined within a cell wall with some other molecules to support cellular functioning. They got nowhere.

Suppose we simplify this down to the molecular level where we demonstrate that Darwin's chance mutation and survival necessity driven evolution begins to work at that level. Could we not then claim that our problem is solved? If you can believe that Darwinian chance and necessity can evolve consciousness in non-sentient beings it certainly seems plausible (to the believers of Darwinism, at least) that it can evolve a living cell as we know it out of the barebone molecular level through evolution.

In the nineties, attempts were made to produce small and simple replicator molecules in cell-free assays and then to subject them to environmental stresses to mutate and evolve them. The conclusion from these studies are startling as summarized by the biologist Peter Schuster:

- Polynucleotides (polymer chains of nucleotides) including the RNA can replicate itself *in the presence of suitable enzymes* in a test tube, without the confinement of cell walls.

- Related small molecules called oligonucleotides are replicable even without the assistance of proteins or enzymes.

- Molecular replicators have been invented even outside the nucleotide family.

Darwinism is sometimes dismissed as mere tautology: Who survives? The fittest. Who is the fittest? One who survives. But now suppose we take molecular replicability to be a physical measure of fitness in the Darwinian sense opening the door for evolutionary processes to be described and analyzed using the language of physics. Since the definition of fitness in terms of molecular replicability is given independently from survival, no tautology is involved.

Schuster also claims that the usual small probability argument against Darwinian mechanism of evolution is negated by evolution experiments of Spiegelman with the RNA molecules. The negation is based on the experimental proof that target oriented adaptation indeed is found to be feasible.

But a closer examination of the results should take the hoopla away. One persistent problem still remains: enzymes (usually proteins) seem to be needed before replicable RNA strands can be made in a test tube. Then it is the same old problem: how did the enzyme come about in the nonliving world?

A second more general problem is this: That molecules which can replicate, mutate, and evolve in the test tube have been produced means that human consciousness has managed to establish a physico-chemical basis for self-replication with the help of enzyme catalysts. But this is very far from what needs to be proven—that inanimate nature of primordial earth can do the same thing through natural processes like climate change without help from intelligent sentient beings, nonmaterial causal agents. It will not do to say that we are part of inanimate nature, too. We are not, we are animate, conscious.

Furthermore, we should not be too hasty to think that solving Darwin's tautology is a big breakthrough, because this so-called breakthrough also exposes a big weakness of Darwin's theory looked upon as a purely materialist theory. In that view, Darwin's theory of evolution through chance and necessity cannot be a genuine biological law. Instead, it must be a corollary of a law of physics of self-replicable molecules. Suppose we state Darwin's theory this way:

Self-replicable molecules that are the fittest adapt best to environmental stresses such as climate change.

In this form not only there is no tautology, there is not even reference to survival. But then can we generalize this law to lead to the evolution of a single living cell, to a conglomerate of living cells, all the way to an organism with defined functioning organs? We cannot even be sure that in this form, Darwin's theory can even pass the first hurdle, applicability to a single living cell, let alone the rest of the hurdles. It is a long way from self-replicable molecules to a cell of many molecules with specific distinguishing structures and purposive functions including the capacity of distinguishing itself from the environment. Replication is not the same thing as reproduction.

Meanwhile we do have the virus. There are conspiracy theories today; people worry if viruses can be made in the laboratory. A virus is an RNA or DNA strand with a protein coating: it is basically a protein without adequate nucleic acid of its own for reproduction for which it needs a host. The conclusion must be that the virus is not a life form, it is not alive. Could it be made in the lab from scratch from nonlife? The answer remains a no, until scientists demonstrate that both nucleic acid and enzymes (usually proteins) can be synthesized in the laboratory from scratch. Of course, this still leaves the door open for the production of viruses via genetic engineering.

In quantum science, the virus, as a part of the nonliving world remains in potentiality until the living comes into the picture, until the virus finds a host.

The verdict is clear: molecular biology cannot provide an explanatory principle for life and its evolution. Propounders of this paradigm should give up their pretense that it is a science life and concentrate on discovering a genuine biology of life and evolution. Sooner or later, they will find that quantum biology removes all the paradigmatic difficulties molecular biology has. Quantum biology also leads to an integration of Darwinism and Lamarckism. Finally, use of quantum biology will redefine the *field of bio-engineering* and remove all the controversy surrounding it.

The final conclusions of this section to convince yourself and take with you are these:

> *Self-replicating RNA chains has been produced in the lab in a test tube with a suitable chemical mixture along with enzymes from a living body; note however: a) this self-replicating RNA is still nonliving; b) nobody has demonstrated the laboratory production of an enzyme.*

> *We cannot use the self-replicating molecules as the starting point, turn on Darwin's chance and necessity mechanism, and produce evolving life;*

> *However, the scientific efforts in this direction are useful to put to rest conspiracy theories about making life in the laboratory, at least viruses.*

> *Biologists must recognize the paradigmatic foundational dicculties of current biology and welcome the paradigm shift to quantum biology within the primacy of consciousness.*

CHAPTER 5

Why and How Evolution

I once had an interesting discussion with a prominent biologist. This was when I was just beginning to have a curiosity in the issues involved in biology and biological evolution. When I asked him, What is the defining characteristic of life? he surprised me. He said, "Evolution. The fact that life evolves."

Why does life evolve? The basic materialist argument goes like this: It's the immensity of the geological time through which life had to pass through. The environment changes, sometimes drastically during the various geological epochs. Life had to adjust to that, adapt to that, survive in spite of all these upheavals. Life does all that through evolution.

But the answer does not entirely satisfy. All environmental change does is to produce a challenge for the organism to adapt. So yes; the organism would certainly change because of all these adaptations. But because the changes are continuous for macro material objects, Newton says and Darwin goes along, one little incremental change at a time, we can always trace the basic features of the original organism in spite of the adaptive changes. And indeed, we have fossil data to show that creatures of the old have adapted their way toward what they are today.

However, there are also new species. Evolution refers to evolution of new species. What explains speciation?

Charles Darwin, the most famous biologist that ever lived believed that the same mechanism that leads to adaptation leads to speciation. His theory of evolution in today's language (Neo Darwinism) has two components: 1) genetic variation; genes are the hereditary components of biological being. The variations come about entirely by chance. 2) Natural selection. The necessity of survival is the criterion nature uses to select among all the variations. In short, Darwin's theory says, evolution is a play of chance and necessity.

Why evolution? The how of evolution, survival necessity, also answers the why question in this way of thinking.

But it did not satisfy me. From my reading of Darwin, I assume it did not even satisfy Darwin. When you think about it, this view won't satisfy you either. In this view, we, even humans, have all evolved in incremental small steps continuously from that one original single cell creature that somehow came about on earth some 3.8 billion years ago driven only by blind chance and survival necessity. No meaning and purpose, no preferential direction! If that's all there is to evolution, our life, human life would not have any meaning and purpose either.

If consciousness is the ground of all being, not matter, then the theory of evolution must be an evolution of consciousness in manifestation. In that model, why and how of evolution bringing meaning and purpose can be explained.

Evolution within Consciousness: the Great Chain of Being

To the spiritual traditions, the ways of experiencing and the resultant bodies of consciousness look like a progression (fig. 2):

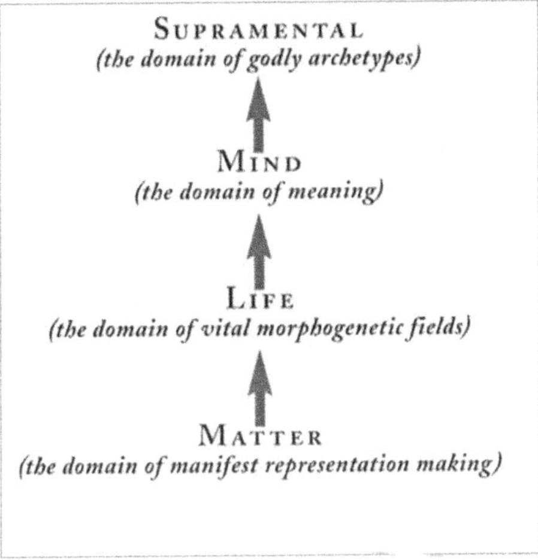

Fig. 2. The great chain of being

This is what the philosopher Ken Wilber calls the great chain of being. Biologists of scientific materialism denigrate the great chain of being. They see life, mind and consciousness, and godliness, all as adaptive epiphenomena of matter that evolution gave us with no consequences other than survival benefit.

I hope the discussion in the previous chapters has helped you to arrive at some healthy skepticism about the materialist premise that life, mind, and consciousness can evolve in matter without the help of other organizing principles. But then your curiosity about the puzzle of the progression above must be addressed.

If science is based on the primacy of consciousness, then the important question is, if consciousness is the ground of all being, the beginning point alpha for evolution, then how does evolution proceed? The answer is enlightening. It was first given by philosopher Sri Aurobindo in 1939, and the Jesuit priest Teilhard de Chardin1(961). Later the philosopher Ken Wilber (1981) and myself (1994, 1996, 2008) elaborated upon the idea. I will now fully discuss the latest version that includes the earlier work.

Note that if you want to stop a questioner from going on asking "What was before that" is to say that ultimate reality has no beginning and no end; ultimate reality is eternal (unqualified nondualism). There is no scope of science (which is a qualification) here. So, our discussion begins with qualified nondualism: consciousness is the ground of being with quantum science to guide the movements of its possibilities.

Aurobindo thought that when evolution is looked upon from the context of the primacy of consciousness, involution—progressive imposition of limitation and forgetfulness—must precede evolution. In the quantum model, the only involution we need is the idea that the unqualified nondual consciousness gives way to a qualified nondual consciousness—the qualification is quantum science.

People from the religion side find it a little hard to comprehend why an "omnipotent" God would play the game of creation within rules. But a play without rules is no fun for us; and we are made in God's image, the purpose of our lives is to bring that image to fruition. So why should God's wanting to play the game within rules be so surprising? When we start a game, how do we start? We lay out the rules of the game.

Also, from a primacy-of-consciousness point of view it is possible to ask, What is the purpose of evolution in general? Even the question, Why evolution at all? And the answer is easy: evolution is needed for experiencing fully the possibilities of consciousness in manifestation. Or, following a statement of the psychologist Carl Jung., the purpose is to make the unconscious—that of which we are unaware, the domain of potentiality—conscious—that of which are aware, that which we manifestly experience.

Once we recognize that the enfolding and the unfolding of the possibilities of consciousness are a purposive play, we can easily lay out the nature of unconscious processing that precedes evolution.

An American comedian had a Noah routine that is very telling. God speaks to Noah from up above; he is giving Noah instructions for the building of Noah's ark, of course. But the modern Noah depicted by our comedian is a little skeptic, "Who is this, really?"

The voice insists, "This is God." But Noah still cannot believe. "Am I on Candid Camera?" he starts looking for the hidden camera. Now God appears in a convincing form and starts giving serious instructions for building the ark once more. But Noah raises that issue of omnipotence. "If you want this ark so much, why don't you build it yourself?" And God says what any quantum scientist would approve. "Noah, you know I don't work that way."

God has to work through us, God plays within rules—the laws of quantum science. But of course, these rules can be subtle and are subtle, they don't have to be deterministic like Newton's. They are quantum laws that permit a certain amount of choice and creativity!

Why Matter is Different (Manifest as Gross and Objective) from the other Bodies (Manifest as Subtle and Subjective)

As Descartes noted early on, mind (and by implication all our internally experienced bodies that include the vital and the supramental) is without extension, but matter has extension. This means mind is just one thing, whereas matter is reductive, micro makes up the macro. This is so as to make manifestation possible. Let's see how.

In chapter 1, I discussed that consciousness becomes self-referent in us in the act of quantum measurement, in the act of choosing actuality from the quantum possibilities. Notice that the existence of the tangled hierarchy depends crucially on the ability of macro-matter to make memory; only macro-mater can make memory long enough to last a human life. In this way quantum collapse can take place only when matter comes into play.

There is no need to posit that tangled hierarchical quantum measurement can happen only in the brain. As discussed in the last chapter, the concept of tangled hierarchical self-reference allows us to distinguish life and non-life and opens us to a concrete definition of life.

Evolution begins with matter, more precisely with the big bang of which there maybe more than one according to modern cosmologists. But the beginning and subsequent steps remain in the unconscious with unconscious processing of the possibilities until the tangled hierarchy and all other ingredients for manifest life is ready. Then, with the advent of the first living cell, life begins as well as the universe, via delayed choice. The next question is, What are the stages of life's evolution?

The stages of Evolution

The script and progression of evolution is quite different in the primacy of consciousness view. What evolves? In the new paradigm view, matter doesn't evolve life but the whole material universe *evolves in possibility* until the first living cell and its environment is ready to manifest the rudimentary biological functions (reproduction and maintenance including waste elimination) and to carry out tangled hierarchical quantum measurement. The evolutionary journey of life now begins consisting of the evolution of the representation-making capacity in matter of the vital morpholiturgical fields.

The evolution from simplicity to complexity that we observe in the evolution of life can now be more fully explicated. Complexity is required to a) make more and more sophisticated representation of the already represented morpholiturgical fields. The purpose is to make room for b) making representations for previously unrepresented morpholiturgical fields in potentiality that correspond to "higher" biological functions to carry out higher needs. Clearly this last item also creates more complexity but of a qualitatively different nature.

With an understanding of the evolution toward complexity the biological arrow of time is no longer a mystery. The more sophisticated an organism gets, the newer and newer morphogenetic/liturgical fields are represented in it in more and more sophistication, the more sophisticated the organism becomes in processing feeling and cognizing via feeling. All this creation of complexity, increasing order and sophistication, takes creativity from consciousness—both unmanifest and manifest.

While life's arrow of time consists of developing more order, more complexity, how does life's environment evolve? When there is no creative demand, consciousness chooses according to the probabilistic laws of quantum physics. This produces increasing randomness and entropy. Therefore, environment's entropy continues to increase. How is life sustained then against this prevailing current of increasing entropy? Life can only flourish in the presence of a large source of negative entropy called negentropy or order. For us, this large source of negentropy is the sun.

Eventually, when the organ brain's neocortex evolves, the organism develops the capacity of representing the meaning processing mind. This begins a new era of the evolution; we have to add the mental representation making capacity to the vital; in. other words, this is second phase of emotions. It has been quite extensively codified in anthropology and sociology as the garden agricultural era. Over time, meaning processing does become more and more sophisticated and more and more devoid of emotion. Right now, we are in the midst of this third phase of mental/rational era of evolution.

Taking note of this third phase of the biological arrow of time, we now can easily project what the future of evolution must hold for us.

The Future of Evolution

In Neo-Darwinism, talking about the future of evolution is a fruitless task. The following lines from the romantic children's fantasy writer C. S. Lewis reflects this sentiment just about perfectly:

> Lead us, Evolution, lead us,
> Up the future's endless stair;

> Chop us, change us, prod us, weed us.
> For stagnation is despair:
> Groping, guessing, yet progressing,
> Lead us nobody knows where.

The reason *nobody knows where* is this: in the Darwinian two-step mechanism of chance and necessity, there is no progressivity, no tendency toward increasing complexity, all is directed toward survivability. This latter primarily depends on fecundity, how much progeny an organism leaves behind rather than much of anything else.

But in our model of evolution, one more phase of the evolutionary arrow of time can be easily predicted. There must be a fourth phase consisting of the evolution of consciousness's representation-making capacity in matter of the supramental archetypes. Since matter as we know it cannot make direct representations of the archetype, we have to do it by the mental-vital route of the archetypal representation making—this is what we call soul making. This then explains Wilber's great chain of being.

Take these important messages of the quantum science of life's evolution with you then:

> *The purpose of evolution is to make the unconscious possibilities conscious, and manifest as embodiment.*
>
> *Quantum Evolution takes place in four phases: i) evolution of the embodiment of the vital, ii) evolution of the embodiment of the vital mixed with mental; iii) evolution of the embodiment of the mental and iv) evolution of the embodiment of the supramental.*

As you can see, this is quite similar to Aurobindo's theory.

Toward the Omega Point

Wilber's great chain of being also has a fourth stage: the spirit. What is that about? Can there be evolution in the spirit phase?

A related question. Can we ever represent the archetypes of the supramental directly in the physical without the help of the intermediaries, the blueprints, the mind and the vital?

I would like to tell you about the work of the visionary Jesuit monk/biologist Teilhard de Chardin. Everybody knows about the biosphere—the web of organic life around and about the material planet earth. De Chardin saw the second phase of life's evolutionary arrow of time as a noosphere, sphere of the evolving mind of humanity. The noosphere is not to be found just in the outer dimension of the world, in our civilization, in our books and in the Internet. There is also an internal dimension of the noosphere.

Teilhard envisioned that when this inner dimension of the noosphere, right now quite fragmented, fuses somehow taking on an internal dynamics of its own, then there will be a new phase in our evolution, a quantum leap that he called the omega point.

Sri Aurobindo had the same vision. He saw the human mind going through a number of evolutionary phases, still all brain based. But then with the development of what he called the overmind, the brain-based humans reach a pinnacle. At the next stage superhuman beings evolve who are able to manifest the godly qualities, the supramental-supravital archetypes of the mind, directly in matter. "Just as the animals are the laboratory for the human, the human is the laboratory for the superhuman," said he.

In the schema that I have presented, the vision of both de Chardin and Aurobindo will fully come to fruition if and when evolution provides for us vehicles of matter capable of representing supramental archetypes directly. This is the glorious future of our evolution.

Think about what it means. Today, some of us take quantum leaps of creativity to glimpse the supramental. Then we make vital and mental representations of the insights gained from them. And then if we are so motivated, we use these new vital representations to elevate organ functions for noble feelings; we also manifest these mental representations to make circuits in our brain. These modes of soul living become habits of transformed behavior of supramental intelligence.

We can also cultivate living in an intuitive mind—live the archetypes depending on moment-to-moment intuition as far as possible. We call such transformed people who live this way saints and sages.

But even if we live by intuitions, even intuitions are glimpsed only through the mind and feelings. It is like drawing a picture from someone else's description. In this way, one sage's soul-living differs from another's enough to cause confusion among the followers especially when the teachings of these sages are translated by lesser people who haven't bothered to transform their mind to the soul stage. Thus, we create religions and differences in our views of God and godliness. These religions fight among themselves and also with science. This has been our big problem throughout human history.

But suppose a direct physical representation of the archetypes were available for everyone to check as needed removing all necessity to fight about whose mental or vital representation of soul-building is more accurate than another's. Also, if we had the benefit of this privileged view, then we could teach others about the supramental qualities just as today we can teach our students mental concepts of our insights without their having to take quantum leaps every time and manifesting the insight.

Earthly matter as we know it cannot do it! Is there any other kind of matter? There is the recently discovered dark matter. Unfortunately, we know so little about dark matter that it is too early to speculate on that basis. And we won't.

CHAPTER 6

Creative Evolution

Let's begin by listing Darwin's errors that I mentioned in chapter 1. One error was not Darwin's error; it was the error of the later concept of molecular biology, that molecules themselves must bring survivability which is Darwin's central idea as the criterion for natural selection. Survivability as a basic need for organisms is perfectly justified; it is experientially obvious to us even at the human level. But molecules have survivability is a very poor idea and as discussed in chapter 4, if we try to build Darwin's theory on the experimentally verified replicability of certain polymer chains, we get nowhere.

The four errors that are Darwin's own are the following:

1. Darwin suggested all evolution is slow and continuous; in this he was following Newton of course with the assumption of continuity.

2. Darwin chose to ignore any effect of development on evolution. There was an earlier theory by Lamarck which theorized that development is the cause of evolution; species acquire characteristics and these characteristics become hereditary somehow. There were a lot of evidence in support of development affecting evolution, but Darwin was not able to accommodate Lamarck's ideas within his theory, not that he did not try. A sort of what we can see as forced error.

3. Ignoring development in evolution is tantamount to ignoring the living part of life; so, Darwin completely ignored concepts of vitalism or vital energy which is a millennia old philosophical theory of what living involves and has been applied in traditional healing practices.

4. The fourth error is he ignored the purposefulness of evolution which should have been obvious even in Darwin's time. Again, this was probably because Darwin was influenced by the scientific paradigm of the time, Newtonian physics where all changes are causal and purpose has no place.

I developed the new consciousness-based theory of evolution in two scientific papers published in the 1990s and published a book on it somewhat prematurely called *Creative Evolution* in 2008. That rudimentary theory is now fully developed in this book.

The central point of the theory is that it explains the fossil gaps as quantum leaps of creativity, a verifiable idea with application in many fields.

In quantum physics, both continuous and discontinuous movement is permitted; the latter is called creative insight involving a process consisting of four stages: preparation, incubation, discontinuous insight, and manifestation.

Preparation consists of creating the urgent necessity of creativity in consciousness, developing the burning question and produce seeds of possible answers.

Incubation consists of unconscious processing of the seeds—thoughts that you imagine may be candidates for a solution; in this way, the processing of seeds is tantamount to processing quantum potentialities really because every thought you imagine becomes a wave of potentialities of meaning as soon as you stop imagining.

Insight consists of a discontinuous quantum leap when a gestalt of possibilities reveals the answer. When an electron jumps from one orbit to another in an atom, it does not go through the intermediate space; it takes a quantum leap; a discontinuous insight of creativity is like this; a new meaning of value arises without any algorithmic step-by-step thinking. Similarly, when a macro-scale species transforms to another, it does not go through the thousand and upon thousands of manifest intermediates that Darwin's theory assumes to exist and that are never found. Instead, speciation takes place as a quantum leap when the gestalt of quantum possibilities of gene mutations is ready for expression and is collapsed producing a new organ or a major improvement of an organ for example. This was the central idea already in *Creative Evolution*: evolution takes place both in continuous steps of Darwinism for micro-level changes and quantum leaps for macro-level changes.

Fossil Gaps, Meaning and Purpose, and Darwinists' Explanation or Lack of it

Everybody knows about the fossil gaps. Contrary to great number of biologists' expectations ever since Darwin, the fossil gaps never filled up. The vast majority of them are real. So, what do they signify? What do they prove?

The Neo-Darwinists, and majority of biologists fall into this category, still insist, they mean nothing. The Neo-Darwinists are sold on a promissory evolutionism—eventually the fossil gaps will fill up. Meanwhile they make a big case every time some new fossil shows up that could be construed as a candidate for an intermediate.

Clever theorists staying within Darwinism have suggested another theory for explaining the fossil gaps: geographical isolation. Suppose a mountain springs up dividing a species in two; one of rather a small population. Small populations produce rapid gene mutations and a lot of quick changes leading to a fast tempo of evolution. Then, the two groups would unite once again and it would look like there has been macroevolution and discontinuous change.

In the same vein, there are clever followers of creation of life in the style of Biblical Genesis—creationism; their idea is that God created life as it says in the genesis, all at once, there is no evolution. The creationists metaphorize all the climate catastrophes that contribute to evolution into one big catastrophe of the great flood and then weaves the story of Noah's ark as the explanation of how living specis survive. How is the theory of geographical isolation different from the creationists' explanation of how some species have survived all the geological catastrophes of the past?

You have to watch these theories; and see that these ideas are not science since they are not verifiable. At least the idea of the great flood is falsifiable; the idea of geographical isolation is not even falsifiable.

It is easy to falsify creationism in a fool-proof way. The Biblical account of the creation of the world and life in six days mere 6000 years ago is just plain wrong; the geological and radioactive timing data is undisputable evidence against it. There is also too much evidence that species evolve from species, we have too much common with monkeys, monkeys have too much in common with lower (in the evolutionary ladder) mammals, and so forth.

The Darwinists got this one right! Later species evolve from earlier ancestors. But Darwinists are dead wrong when they claim all speciation is continuous, that human being developed from monkeys in a gradual continuous way. The fossil gaps say otherwise, macroevolution happens via a quantum leap, often more than one.

There is also the question of meaning and purpose. Darwinists are dead wrong when they say there is no meaning and purpose to evolution, there is no play of intelligence in the origin and evolution of life, and there are no "lower" creatures that serve the lower purpose of mere survival and "higher" creatures that serve "higher" purpose such as the exploration of archetypes.

For meaning to evolve as an adaptive survival value, matter must be able to process meaning. But recent research has shown that computers, being symbol processing machines, can never process meaning ambiguity from scratch; there is a category difference similar to that in grammar between syntax and semantics.

How can nature select a quality that matter cannot process and therefore cannot offer for selection? It can't.

How does our ability to discover a scientific law arise? We can be specific. Consider Darwin's own law of survival of the fittest which as I mentioned many times, has some grains of truth in it. The discovery of such a law itself has survival value; that is not the question here. The question is, can the knowledge of this and other scientific laws be coded in matter somehow? Can they arise from the random motion of matter somehow? Attempts to prove that such is the case have not had any success whatsoever.

The question of how consciousness evolves is another case in point. To the question, can matter ever codify consciousness? there is that hard question: how can interacting objects ever produce a subject-object split awareness? If material interactions can never produce consciousness, to think of consciousness as an adaptive evolved value does not make any sense.

The conclusive scientific proof that there is purpose in Consciousness's creation is that there is a biological arrow of time. As previously discussed, the purpose of evolution is to create complexity and the biological time's arrow is directed from simplicity to complexity of living organisms.

Two Tempos of Evolution, Punctuated Equilibrium, Hopeful Monsters, and Neutral Genes

There are maverick biologists, who from time to time have suggested that the fossil gaps signify that apart from the slow tempo of evolution that Darwin proposed and Neo-Darwinists are sold on, there is a fast tempo of evolution—so fast that there isn't time to lay down fossils. In other words, evolution is like punctuated prose; there are abrupt and discontinuous punctuation marks among the otherwise continuous prose. The idea is called the theory of punctuated equilibrium (Eldredge and Gould, 1972; Gould, 1980). Unfortunately, the proponents of the theory were unable to find a mechanism for a fast tempo. Scientists don't like living in an explanatory vacuum: If no viable theory of fast tempo is available, let's proclaim that Darwin's is the right theory of all evolution and explain away the fossil gaps, fast tempos, and the problems of development of novel form! This has been the position of most biologists.

Some theorists do seriously wonder how novel traits can come about in evolution since many genetic variations have to be acted upon simultaneously which is against the grain of Darwinism. There is the idea of "hopeful monsters" of biologist Richard Gold-schmidt (1952) —a whole bunch of new gene variations find expression all of a sudden. And furthermore, there is the idea of neutral gene: the required gene variations remain dormant because they are neutral with respect to survival value until their totality adds up to a significant novel form (Kimura, 1953).

There is also the work of biologists who point out another important piece of data suggesting discontinuity. Before all great creative evolutionary epoch of macroevolution, these theorists point out, that there always occurs some kind of catastrophe leading to a massive extinction of biological species. These catastrophes clear up the biological landscape for new evolution of species. And so, the new evolved species have no need to compete for survival.

Now here's what all this suggests. The gaps in the fossil data are some of the best proofs of the existence of quantum collapse of downward causation, which are discontinuous as pointed out by von Neuman. These are also the creative acts of conscious intervention. Creativity occurs through quantum leaps, taking no time. Here is the new mechanism for the fast tempo of evolution! Let's see how this one idea integrates the thinking of practically everyone mentioned above: Of the punctuation theorist, because indeed it takes creativity, one giant discontinuous leap to manifest all the right genetic possibilities for the raw material for making a new form.

Amazingly this quantum theory of creative evolution also incorporates both the ideas of hopeful monsters and neutral gene in one fell swoop (see below). It also satisfies the catastrophe thinkers because catastrophes are part of creativity, destruction before creation. The destruction is needed to open up new ground for the play of the new. The appropriate metaphor for quantum consciousness in this aspect of creativity is Siva's dance (fig. 3). Consciousness in this special aspect of Siva, the king of the dancers is first a destroyer and then a creator.

Fig. 3 Shiva's Dance: You can see both destructing and restructuring in the symbology. The right foot stomps on the bottom figure of ignorance signifying destructing of the existing knowledge and the uplifted left foot signifies restructuring.

Unconscious Processing and Quantum Leaping: Creativity in Evolution

The biggest problem of biological macroevolution is this: a macro-evolutionary step requires so many changes at the genetic level, so many mutations! To mention it for the umpteenth time, the development of an eye from scratch requires literally thousands and thousands of new genes. But each gene variation, according to Darwinism, is selected individually. The likelihood of its being beneficial is quite small; in fact, gene variations are most often just the opposite. Chances are high that individual selection would eliminate most if not all such gene variations.

Think of a specific problem, the evolution of the long neck of the giraffe (Hitchins, 1982). According to Neo-Darwinists, the long neck happened through slow accumulated variations and their selection, gradually leading to longer and longer necks, enabling giraffes to reach higher and higher leaves of trees where the food is in ample abundance. But this is too simplistic. Because it is a fact that developing longer neck vertebrae require many concurrent modifications.

It is worthwhile enumerating the details. As the vertebrae become longer, the head must become smaller because it is more difficult to sustain it atop a long neck. By the same token, the circulatory system has to produce higher blood pressure; valves must originate in order to prevent overpressure when the giraffe stoops to get a drink. The lung size has to increase so that the animal can breathe through a much longer pipe. Additionally, many muscles, tendons, and bones have to change harmoniously in synchrony; in fact, the entire skeletal frame has to be restructured in order to accommodate lengthened forelegs. It goes on and on. Clearly, much more than neck-lengthening gene mutations have to be involved and with what amazing coordination! All this through cumulative chance? All thisonly through local material interactions? Not credible, is it?

The idea of unconscious processing followed by quantum leap gives us relief from such puzzles of Darwinism. All the genetic changes in an existing organism are quantum possibilities; that gene mutations are quantum events was first pointed out by the physicist Walter Elsasser (1981). Consciousness acts on them and simultaneously uses suitable quantum morpholiturgical fields and psychophysical parallelism for guidance on the form. Only nonlocal psychophysical parallelism can permit such nonlocal coordination between many simultaneous organ developments.

When a coordinated gestalt is completed in potentiality for a fully functioning change, consciousness chooses; the result is a new species with new organ(s)—for survival and/or for new ways of exploring the environment and the archetypes.

Maxwell's Demon and the Entropy Question

Of all the physical laws of pre-quantum physics, the entropy law is the most intriguing. Entropy is disorder. When a system is in a state of maximum disorder such as the air environment around us or the vast oceans of water, you cannot extract any of the vast amount of heat energy of random movement of molecules locked up in those systems. This is the purport of the entropy law. What a wastage! So, physicists play with ideas of somehow getting some order out of a system in disorder; Maxwell's demon is one such idea.

Imagine, following the physicist Clerk Maxwell, that a demon is sorting out slow molecules from fast molecules placed in a box, thus creating order in an ensemble of molecules that is initially all randomly mixed up so far as speed is concerned, a situation, physicists call thermal equilibrium. The demon puts a partition with a trap door in the middle of the box of molecules and sits at the trap door (fig. 4). Whenever the demon sees a fast molecule at the trap door moving from left to right it lets it go through the trap door. And whenever a slow molecule is ready to pass through the trap door from right to left again the demon lets it through. So, by his trap door trick, the demon has separated slow and fast molecules, has created order from disorder.

This seems to contradict the entropy law: in natural processes, entropy, disorder always increases or at best, stays the same. The physicist Clerk Maxwell thought up this demon in order to demonstrate a possible violation of the entropy law. However, another physicist, Leon Brillouin pointed out how the entropy law is saved. Every time, said Brillouin, the demon looks to see if a molecule is the right one to let through the trap door, it has to spend energy to get that information, and that costs entropy, so entropy increases overall.

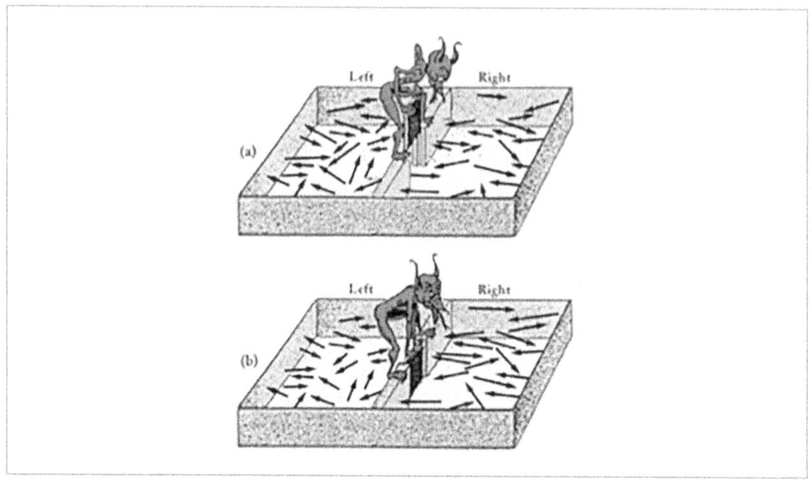

Fig. 4 Maxwell's demon. (A) The molecules of both chambers are disorderly, Both slow (Short arrow) and fast (long arrow) molecules. (b) The demon has produced order in both chambers by his trap door trick; the left chambers now has only slow molecules;

So, the evolution of life, order from disorder, does not have to be in violation of the entropy law. Still, it is bothersome. For Darwin's process of natural selection to work for speciation, thousands of gene mutations have to accumulate without being selected again (the neutral gene hypothesis). To sort this out would require a demon testing each gene mutation for neutrality and survival benefit. This would cost the environment a lot of entropy.

Darwin's theory can be attacked on this score alone because I don't think there is any definitive evidence that such entropy increase has taken place in life's earthly environment. In fact, the Nobel laureate Prigogine and others did not think so and this is why they tried to devise theories of evolution based on an entropy-law violating dynamics from the get-go (valid for so-called thermodynamically open systems) that produces order within chaos. Unfortunately for scientific materialism, those ideas don't work either.

All this is still debatable. What is not debatable is that a lot of time must elapse to account for that much entropy increase. Darwinian evolution has to be very slow!

Now the quintessential question. Doesn't the same criticism apply to conscious processing of the genetic variations? In fact, can we not look upon the Maxwell's demon as a conscious processor? Then do we not have the same entropy problem facing us?

For creative evolution, the situation is saved by unconscious processing. It is conscious processing that costs entropy and time. But in quantum theorizing, the gene variations are quantum possibilities. Biologists do Newtonian thinking assuming that the gene variations would collapse as soon as they arise without any help from consciousness. But we know better: quantum collapse requires consciousness and its power of "downward causation" in the form of choice. And any gene variation that cannot be expressed in creating a macroscopic trait remains uncollapsed until it can. Consciousness does not collapse the unexpressed gene mutations—quantum possibilities all—until a whole configuration of them when expressed makes a new form. Consciousness waits for the right moment, as we are found to do in our own creative process.

The previously mentioned idea of neutral gene is a wonderful idea after all, except that without the use of quantum theory the idea falls prey to Maxwell's demon paradox. In the quantum view, unexpressed gene variations are naturally neutral with respect to selection because they are not manifest; they are only possibilities.

And Goldschmid also had a sneak intuition of quantum creativity in evolution when he conceived the hopeful monster. Produced by chance a hopeful monster would be just that, a hopeful monster with no chance to continue; to produce a survivable new life form by chance is practically an impossibility. But when produced as a result of unconscious processing and quantum leap akin to creative insight, the idea becomes credible.

What is crucial also is that consciousness has the vital blueprints—the morpholiturgical fields—of the novel form as well as its function (physiology) in potentiality giving it a rough guideline of what to look for that will embody a conscious idea of purpose. When does consciousness choose? As you recall, consciousness needs amplification before it can choose. Therefore, collapse does not take place at the micro genetic level. An amplification of the micro genotype to the macro phenotype must take place in possibility. When there is a match between the macro-physical form and physiology and the vital blueprints, call the match a morpholiturgical resonance (following Rupert Sheldrake's idea of morphic resonance), collapse of the possibility waves precipitates, a quantum leap takes place all at once and consciousness has succeeded in making a physical trait or organ of new function (hardware with the appropriate form and physiology), a representation of new possibilities of conscious exploration. There is no fossil record for the intermediate stages because there are no intermediate stages!

Darwinists are like magicians who have the ability to present things in such a way as to portray the illusion of gradualism. People joke about it:

There are two kinds of people: the first kind can deduce profound conclusions from very incomplete data and the second kind who do not have this ability. The implication is that biologists belong to the first kind.

The biologist Gavin de beer claims that the evolution of the eye is gradual because it goes through progressive stages. The first stage shows an organism with a light-sensitive spot and a primitive lens. In the second stage something like the eye of a jellyfish evolves: it has the lens and light sensitive cells forming a cup-shaped retina. In the third stage, for example in the eye of a tadpole not only the lens and light sensitive cells forming the retina are there but now the retina is formed from the lining of the brain cavity. And of course, this is followed up by still higher stages when the eye is backed up first with the mammalian brain and subsequently with a brain with a neocortex. Gradual evolution? Hardly. The evolution from scratch of a primitive light sensitive spot and a lens requires many coordinated genetic changes and it takes a quantum leap to achieve them. And the evolution of retina with the capacity of brain processing is a still bigger quantum leap requiring even many more genetic variations. Etc., etc.

Creationists try to make a big case against evolution raising the entropy question: Sun's negentropy may provide the order that life displays as a whole, but is there evidence for the large amount of entropy increase of life's environment that evolution must produce? The paradox of Maxwell's demon goes into the heart of the question and identifies that if the processing of information were conscious, then indeed, there is an entropy increase problem for which there is little evidence. Quantum science of evolution resolves this entropy puzzle brilliantly: the processing of information in creative evolution is mostly unconscious, taking place in potentiality, costing no entropy. Take this idea with you.

Are there Intermediates: The Archaeopteryx Controversy

Some Darwinists know in their heart that geographical isolation theory of fossil gaps is pretend science because it is unverifiable. So, they continue to deny the existence of fossil gaps by demonstrating that there are intermediaries between one macro species and another; in this way there is a continuity.

There are indeed a few cases that Darwinists hail as clear evidence for intermediates required by the gradual theory of evolution—Darwinism. Let's examine one famous case of a so-called intermediate—the archaeopteryx. The case of the archaeopteryx is well known to many people because the comic strip *B. C.* regularly makes fun of it.

When the archaeopteryx fossil was discovered, it was immediately hailed by the Darwinists as an intermediate between reptiles and birds because this species seems to have fully developed feathers as well as clear reptilian features of long tooth-equipped jaws and trailing tail. Unfortunately, the creature lacks the strong sternum to anchor the powerful muscles that are required to operate the wings in birds. Therefore, some researchers have concluded that archaeopteryx did not fly.

But then what are the feathers for? In Darwinism, where every feature evolved has to be an adapted feature with survival value, this becomes an embarrassment! This "preadaptation" is an anomaly that haunts Darwinists—how can an adapted feature appear before it is used? So, Darwinists come up with the suggestion that the feathers of archaeopteryx are not for flying but for conserving heat overlooking the fact that the feathers found on archaeopteryx were flight feathers not the down-like heat conserving feathers.

The theory of creative evolution has a very simple answer for all cases of such preadaptation. It is that, why deny it? The pre-adapted form is indeed arrived at via a quantum leap, but it takes further adjustments to make the creative trait gained useful for the purpose it was designed. There are many cases of human creativity where the final discovery had to wait for many more pre-discoveries—quantum leaps in their own right—to take place first. The case of the discovery of the quantum math by Heisenberg and Schrödinger in 1925-26 is a wonderful case in point: Planck's discovery of the quantum in 1900, Einstein's discovery of the idea of wave-particle duality of light in 1905, Bohr's discovery of quantum leaps in atoms in 1913, de Broglie's discovery of matter waves in 1923, all these were genuine cases of pre-discoveries: their significance was obscured by paradoxes, quantum physics could not fly, without the compete discovery of the quantum equation of motion by Werner Heisenberg and Erwin Schrödinger.

In other words, let's face it! an archaeopteryx *is* a bird, but a bird in progress, with shortcomings. Indeed, many researchers do agree that it should have been able to fly as high as the branch of a tree to save itself from a predator. This is the aspect that the cartoon *B.C.* makes fun about. You can make fun about it too:

Why did the archaeopteryx get the worm?
Because it was an early bird.

To be sure, there had to be evolution beyond the archaeopteryx: a strong sternum, muscles supporting flight, and perhaps even suitable brain power before an organism could fly properly like a modern bird.

In this way you see clearly that the fossil gaps are real, they are the general rule for speciation and evolution, and they are evidence of biological creativity, of quantum leaps in evolution. And as such they provide us with some of the most spectacular evidence for consciousness and the quantum in biology. *The few cases of intermediates are examples of pre-adaptation, a sneak preview of momentous biological creativity.*

One question still needs to be addressed. In human creativity, creativity consists of the individual human's creatively taking a quantum leap from quantum-unconscious (part of the unconscious with previously uncollapsed possibilities) to manifestation; the individual has a clear role to play in the stages of preparation and again in in the final stage of the creative process—manifestation. Clearly, some form of manifest consciousness must have a role to play in the creative stages of preparation and manifestation. In biological creativity of evolution, the role of the individual organism in human creativity is played by species consciousness, sometimes even by a higher group consciousness like the genus, order, even class or phylum. The individual organisms are correlated together as a whole in the unity consciousness of the whole species or group via nonlocality.

Take these two ideas with you:

The few cases of intermediates found "inside" the fossil gap are examples of preadaptation—premature quantum leaps.

Who takes the quantum leap in biological creativity? The entire species or even a bigger group like the phylum.

The Role of the Species or Group Consciousness

In Darwinism or Neo-Darwinism, the organism has no role to play; this is bitterly criticized by organismic biologists who maintain that the development of the organism, in fact the organism itself, must have a role to play.

In the scenario above of how the quantum leap takes place, it is clear that development of an organ does play a crucial role. Now we can enunciate the role of the organism as well as species or group consciousness when we take account of the catastrophes that precede quantum evolution.

For human creativity, all creatives know that creativity requires a motivation, an urgency, usually a burning question. From the point of view of the whole quantum consciousness, there is the motivation of purposive evolution. When an environmental catastrophe occurs, this evolutionary motivation percolates through to the individual organisms in a hurry because it coincides with survival necessity. This synchronous intention in the whole group reflects a strong motivation to change.

This sounds more like Lamarckism, but there is now experimental evidence in favor of this kind of Lamarckism.

In 1988, the biologists John Cairns and his collaborators demonstrated that a certain kind of bacteria, when proffered food (a form of lactose) that it cannot directly digest, but can digest if it undergoes a one-step mutation, hastens up its own mutation rate and survives. This phenomenon is called directed mutation. So even in microevolution, Lamarckism sometimes persists.

Biological organisms have nonlocal connections in the vital arena through their quantum morpholiturgical fields mediated by nonlocal consciousness. Because of the dominance of the rational mind over feeling, this nonlocality is currently somewhat obscure for us humans. But the rest of the biological world, being non-mental, at least largely so, is not limited that way. This nonlocal connection through the vital body acts as a group consciousness (think of it as a generalized group ego). It is this group consciousness that intends evolution in response to rapid environmental changes and if this intention is consonant with the purposive movement of quantum consciousness, consciousness goes along with this evolutionary intention.

Nonlocality in Evolution

Is there any direct evidence for nonlocality in evolution? There is. There is a phenomenon called co-evolution where two entirely different species evolve together in order to better survive. Many plants and ants have such a symbiotic relationship. Plants provide the ants with shelter and food, and the ants drive away other harmful insects and animals, and even help keep other vegetation away. We can see co-evolution most easily in such symbiotic relationship. Indeed, it is found that some plants independently evolve seeds with an edible fatty lobe as a reward for ants. The ants on their part reward the plants by carrying away the seeds to their homes, eating the fatty lobes, and leaving the seeds to grow in places that are favorable for the plants to grow. The reward dynamics suggests that the archetype of goodness is in play.

The Darwinists argue that each species puts selection pressure on the random variations of the other as part of the other's environment. But this denies the fact that the probability of simultaneously beneficial variations is extremely low. A much more probable answer is that the necessary mutually beneficial variations are held in limbo until the time is opportune for nonlocal consciousness to collapse the two correlated sets of variations in two co-evolving different species simultaneously. This is a variation of what Carl Jung calls synchronicity.

Synchronicity

There is now some consensus that the dinosaur extinction some sixty-five million years ago was brought about mainly by a large meteor shower. This made room for the very important evolution of the mammals that eventually led to the evolution of the human being. Evidence shows that mammals were already on the scene, but the catastrophe made room for their creative expansion.

So, is the evolution of humans on earth brought about by pure meaningless chance? If that is so, then how can we uphold purposiveness of biological evolution, when clearly, the purposiveness of consciousness needs the help of a chance event?

There is no need to see contradiction here with the scenario of biological creativity and purposiveness. Chance contingencies are often very important in the history of a creative act, except that we see them as components of events of synchronicity. An event outside in the material arena (the meteor shower) and an event inside in the biological arena (the act of biological creativity) occur simultaneously and meaning and purpose emerge in the prolific evolution of the mammalian class. This is synchronicity.

In fact, these events of synchronicity are important because they open up the evolutionary landscape for the newly created macro-organisms. They also create a sense of survival/urgency for evolution in the organisms that survive the catastrophe. A sudden change of environment requires an equally sudden evolutionary jump. There is no time for waiting for slow Darwinian evolution to bring adaptation.

Ideas to carry with you:

Catastrophes, big geologic al events of extinction, provide the motivation of creativity of the surviving species or groups.

Catastrophes are not events of pure chance; instead, they must be looked upon as synchronicities.

Creativity and Complexity: Why biological Creativity in the Evolution of Life is Progressive

I have mentioned the adage before, God (Oneness consciousness) creates humans by His (or Her or Its) own image. By looking at our own creativity, we can gain insight as to how God's creativity in evolution works.

I have never been to Bali, but I have always been fascinated about authors writing about the creative culture of the Balinese. These authors portray a wonderfully innocent and "unsophisticated" culture that is ever so creative. But guess what? Like their culture, the Balinese creativity also remains undeveloped and unsophisticated. The Balinese go on "inventing the wheel" over and over again, so to speak.

There is a reason for creating sophisticated cultures or "civilizations" after all! That is the way that we open the door for the discovery of sophisticated concepts like scientific laws or the complex tapestry of a musical symphony. Sophistication requires creativity to proceed in a ladder-like development. The contexts discovered earlier have to act as scaffolding for creative acts leading to more complex expression later. Creativity itself evolves as the context of its invocation evolves.

What does all this mean for biological creativity in evolution? Every act of biological creativity leading to a new trait and a new species is a new expression of the archetypal biological functions. The new trait opens up the possibility of the evolution of further new traits for either a more sophisticated representation of the same biological function or an entirely different biological function.

Some traits work better as scaffolding for new traits giving evolution an enormous opportunity for the increase and proliferation in complexity. Others are like a blind alley; they do not allow much further creativity; for them stasis is the rule.

This is exactly what we find in the evolutionary data. There are the frogs which have existed longer than the mammals and consist of thousands of species. And yet the anatomical similarities between them compel biologists to classify them all in one order. One type of frog remained in stasis to such an extent that ninety-million-year-old fossils and the present-day variety are classified in the same genus, Xenopus. Then there are the mammals, a much younger group compared to the frogs, but depict such an enormous scope for fundamental creativity that zoologists classify them in 24 orders. Imagine the corresponding rapid rise of complexity.

Take this very important idea with you about not only biological creativity but all matters of creativity: *creativity progresses in a ladder-like way in progressive stages of more complexity and sophistication.*

Situational Creativity and Stasis

In between the quantum leaps of quantum evolution, what happens? It is easy to suspect that Darwinian slow mechanism should be enough to cope with slow environmental changes. But there are subtleties.

First note that the creative leaps express a whole bunch of new gene mutations, new genes. In some combination, these genes make specific organs. But a gene can be used and is used in more than one combination and in more than one context. In this way, you can easily see that the creative leaps of evolution cumulatively build up what is called the gene pool of the entire species, a pool that now can meet the adaptive needs of the species without having to develop new genes.

In human creativity, this ability to adapt to societal needs by inventing new combinations of the old already discovered ideas is called situational creativity or invention as opposed to fundamental creativity of discovery.

Why is invention or situational creativity easier than discovery or fundamental creativity? Once an idea is expressed a few times, the probability for collapsing it subsequently increases something akin to the phenomenon of conditioning in psychology.

For the biological situation, a good example is the famous case of gypsy moth in London that underwent a change in color to brown in response to environmental pollution. The "brown gene" was already in the gene pool. The usual Darwinian thinking goes like this: The individual moths that were born with this "brown gene" have selective advantage over the moths of the old color. Hence, they survived, whereas the others didn't. So rapidly, natural selection wiped out the moths of the old color in favor of the new brown.

But there are problems with blanketly using this kind of thinking. The author Gordon Rattray Taylor (1983) points this out. ". . . it (the genome) must have a facility for storing. . .potential forms and for realizing or activating them when circumstances demand a change." It is probably okay to ignore the storage problem in the case of change of color or size. But if more than a few genes are involved in a change, it is hard to see an easy solution for the storage problem that Taylor is alluding to within scientific materialism.

But a biology-within-consciousness depiction of what happened in the gypsy moth case would go like this: threatened by the polluted environment, the group consciousness felt the urgent need for change. Using situational creativity, it was able to invent the right combination of existing genes giving rise to the appropriate macrolevel change to deal with the environmental challenge (in the gypsy moth case, the change in color).

In this picture the problem of storage of information for potential change that is currently not in use is solved because the information about the genes are literally part of the quantum potentiality (species unconscious) with an added advantage. Once a gene is expressed a few times, the probability for collapsing it increases something akin to the phenomenon of conditioning in psychology.

The advantage of the present approach, the second scenario, should be obvious. When a macro-level trait has a one-to-one correspondence with a single gene, the two scenarios are almost equivalent. This is called the correspondence limit: in certain cases of situational creativity, creative/quantum evolution becomes approximately the Darwinian model of adaptation.

It is a fact that there are only a few cases where there is such one-to-one correspondence between a gene and a trait. More often a combination of genes is needed for a new, even previously adapted trait. And the probability for a natural occurrence of the right combination simultaneously in enough offspring so that the new trait can take root (that would be a hopeful monster in Goldschmidt's terminology) is very low. In this way, the Neo-Darwinists get away with murderously inept pictures taking advantage of the present state of poor knowledge of gene to trait correspondence.

So, the new picture is actually an improvement over the usual neo-Darwinian model of adaptation although the net effect is the same, adaptation. So Neo-Darwinians should be happy.

Finally, as Stephen Gould and others have noted, the fossil data also show vast epochs of virtual stasis in the evolutionary history of all species. This is a serious embarrassment for Darwinism. But for creative evolutionary thinking, this corresponds to our own ego-homeostasis, the limit of conditioned existence when no creativity, situational or fundamental, is needed to deal with environmental changes.

There is an important lesson here for explorers of paradigm shifts— the correspondence principle. Take these two ideas with you:

> *Paradigm shifts never completely replace the old paradigm; the old paradigm is still useful in some limited arena of application, the correspondence limit.*

> *In the case of biological creativity, the Darwinian paradigm is the correspondence limit of the quantum paradigm of evolution.*

In the following section, we will see that Darwinian theory of adaptation applies in the manifestation stage of biological creativity.

A New Role for Natural Selection: The Manifestation Stage of Biological Creativity

What is then the role of natural selection in evolution if it cannot even be said to have produced intra species changes in response to environment as in the case of the gypsy moth? The biologist Ernst Mayr has called natural selection the sculptor of evolution and George Simpson likewise compared natural selection's role to that of a poet or a builder. These biologists were ready to attribute all the power of biological creativity to natural selection and we are taking it away.

Is there any role at all that natural selection plays in evolution? The role is obvious when we consider that the creative process has four stages: preparation, unconscious processing, insight, and manifestation. To recap, in biological creativity, preparation creates the burning question of survival. The stages of incubation and insight we already discussed. What is the manifestation stage of biological creativity in evolution?

It is simple. Now that the new has evolved, it does have to adapt to the environment; otherwise, it would not survive! This creative necessity of environmental adaptation is what Darwin recognized as the necessity of survival. Darwin was a believer of Newton's cause-driven physics and gradualism, so he called the necessity of survival to be driven by the selection pressure of environmental changes.

Instead, the right view is that the selection pressure of environmental changes acts in the preparation stage, contributing to the creative urgency in consciousness. And the necessity of survival is instrumental to drive the manifestation stage of expression from the future; it is purposive. The manifestation to adapt to the environment takes place either through situational creativity (mini quantum leaps) or in a few simple cases through Darwinian mechanism, through what Niles Eldredge called "tracking" the environment or even through Lamarckism—more information gathering and then make the necessary adjustment.

The fossil data actually supports this new interpretation of the role of natural selection. If natural selection was the causal mechanism for adaptation, it is very hard to see how a species can be maladapted to the environment. In the new approach maladaptation can be readily explained as an incompleteness of the creative insight which leads to the inadequacy of the evolved trait or the incompleteness of the manifestation stage of the insight—failure to find an appropriate situational adjustment or the failure to find a niche that fits the trait.

For an example of creative maladaptation in which a species develops a trait that does not quite fit the survival necessity, consider the stinging bee that assures its own death by the very act of stinging. It does not make sense for Darwinism but makes perfect sense as an inadequate creative manifestation. Another example is the peacock's tail that is counterproductive from a utilitarian view and natural selection should not have allowed it. But if evolution is the result of creative quantum leaps, then survival necessity is just one of the agendas for the creator and can be relaxed on occasion. In this occasion, perhaps the agenda was the archetype of beauty.

Darwinism or Creative Evolution? Further data in Support of Biological Creativity in Evolution

Evolution is! But is evolution synonymous with Darwin's theory or is there another theory that better satisfies the theoretical demand for logical consistency as well as the empirical demand for agreement with data? In the last three chapters we have discussed the trials and tribulations of Darwinism/Neo-Darwinism throughout its one hundred and fifty-year history. We also have looked at how creative evolution unambiguously expands the scope of Darwinism and explains some of the major data for evolution for which Darwinism seems ad hoc and ambiguous at best. In this section, I present additional data in support of creative evolution.

One characteristic of fundamental creativity is that when things are primitive and there is an open vista of ideas to discover, the quantum leaps tend to be gigantic. As ideas are explored, the fields of individual creative exploration become narrow. In science, it is much more fun to explore a new field than a field for which there are many known contexts and creative exploration has become canalized.

In biological creativity, creative evolution would then predict that the more we go back in geological time, the more astounding cases of creative leaps in speciation should have taken place giving rise to many new contexts (which translates as biological groups like phyla) for further evolution. And with time, there should be a corresponding reduction in the evolution of new contexts. Eventually, there would only be the narrowest of context for change—species to species within the same genus.

What does the data say? The zoologist James Brough points out that there have not been any new phyla since the Cambrian age 500 million years ago. And new classes within a phylum stopped emerging some 400 million years ago, since the lower Paleozoic era. The next lower hierarchy, new order stopped emerging about 60 million years ago—at the end of the Mesozoic era. To quote Brough,

"Evolution seems to have worked in a series of more and more restricted fields with large scale effects steadily decreasing. . .as to the future, evolution may go on working in smaller and smaller fields until it ceases altogether."

This is exactly what creative evolution predicts except of course, though evolution of physical/vital all but ceases with homo sapiens (human species), evolution of the representation-making of the mind takes off.

Needless to mention, Darwinism favors the opposite trend. If evolution is the accumulation of small variations, we should expect only over long periods forward in time new orders; new classes taking even longer periods forward in time; and new phyla the longer time forward yet.

The maverick Richard Goldschmidt thought not. It is the phylum which contains classes, said he, "Can this mean anything but that the type of the phylum was evolved first and later separated into the type of classes, then into orders, and so on down the line? This natural naïve interpretation of the existing hierarchy of forms actually agrees with the historical facts furnished in paleontology. The phyla existing today can be followed the furthest back into remote geological time. Classes are a little younger, still younger are the orders, and so on until we come to the recent species which appear only in the latest geological epochs."

What is the Darwinian reply to this? The Darwinians' guru Ernst Mayr glibly declared, the categories (phyla etc.) are man-made artifacts.

To repeat, what is observed in biological evolution gives credence to the creativity theory of evolution, not Neo-Darwinism. For the fossil data, new sudden bursts of flora and fauna are called radiation. During the early Cenozoic era fifty million years ago, sometime after the extinction of the dinosaurs, mammals suddenly exhibited an amazing radiation into about 24 different orders—rabbits, rodents, elephants, Cetacea, primates, included. The resulting 'Age of the mammals' took only 12 million years to establish itself. After that the progress slowed down.

It is difficult for Darwinian gradualism to explain this kind of rapid diversity. It goes against the grain of Darwinian thinking.

Another important facet of creative evolution shows up in the phenomenon of parallel evolution the most famous case of which is the evolution of placental (from which humans originated) and marsupial (such as kangaroos) mammals. The similarities of form and function of corresponding animals of the two groups are remarkable. In Darwinism plus geographical isolation model, the genes, although originating from a common ancestor would diversify rapidly changing the organs and their functions as well. But in the creative scenario, the forms and functions are imposed by the morpholiturgical blueprints which remain the same for both parallel lines accounting for the persistence of the similarities of form/functions.

To summarize: Darwin's is a great theory with tremendous historical significance, but it is incomplete in spite of modifications such as Neo-Darwinism. It is most important to make four adjustments all incorporating creativity in evolution to make a complete theory of evolution:

1. *The concept of creative unconscious processing has to be introduced to account for so many missing links; species (groups) change in potentiality until enough change has taken place to make a new organ(s).*

2. *Discontinuous quantum leaps of creativity have to be introduced to explain the fossil gaps of discontinuity in evolution.*

3. *The idea of natural selection is to be replaced by the concept of creative conscious choice in response to not only survival needs but also higher needs leading to big quantum leaps followed by mini quantum leaps. Natural selection takes place only during the manifestation stage of creativity. And finally,*

4. *Purposiveness is to be included (that is, the conscious choice of quantum leap is purposeful) to account for the biological arrow of time—the evolution going from simplicity to complexity.*

Behold! These adjustments cannot be made staying within scientific materialism. You have to rethink evolution within the quantum worldview.

CHAPTER 7

How Lamarckism Works: The Evolution of the Vital Software in Quantum Evo-devo

In the original evolution scenario that Sri Aurobindo developed, evolution proceeds with first, the progressive evolution of the representations of the vital and after that is over, the progressive evolution of the representations of the mental. The drive is for consciousness to know itself in manifestation. When we apply quantum science to this theory it translates thus:

1. Consciousness expresses itself via the manifestation of its quantum potentialities—new genetic potentialities as well as vital potentialities of form and function.

2. What vital potentialities of form and function will be available when new genetic potentialities come into the picture and what will manifest for a given species of organism depends not only on unmanifest consciousness but also on the manifest species and organism. At every homeostasis, continuous slow Darwinian evolution will result unless and until there is urgent burning conscious intention to change at the manifest level.

3. When the intention at the manifest level is in consonance with the evolutionary purpose of the unconscious, quan-

tum leaps of evolution take place.

In this way of looking at the quantum theory of evolution, evolution is initially evolution of the vital, giving rise to newer and newer forms and functions and enabling progressively more and more expressions of consciousness. Also, evolution has two tempos: one slow, continuous, and mechanical as in Darwin's theory; the other is rapid, conscious, and creative, consisting of an instantaneous quantum leap followed by many mini quantum leaps and natural selection in the manifestation stage of creativity.

By and large, the fossil data supports this quantum scenario of evolution as we have seen in the last chapter. There comes a crisis, most often brought about by large scale climate change. A significant portion of the manifest populations respond to the crisis, consciousness chooses to manifest new organs from the available genetic variations waiting in potentiality and quantum leaps occur. A new homeostasis comes about after a brief manifestation stage of biological creativity.

The creative process is the mechanism for the macro evolution of life– macro evolution is an expression of biological creativity. The creative process is not an isolated idea just to explain the evolution data; the same creative process is used by us humans not only for our own creativity in science, the arts, literature, and music, but also for our personal and social transformation.

The idea that evolution is a creative act of species group consciousness is considered as a heretical idea in conventional biology; however, the idea is implicit in Lamarckian theory of evolution. And this idea has now been verified.

Recall please! In 1988, the biologists John Cairns and his collaborators demonstrated that a certain kind of bacteria, when proffered food that it cannot directly digest, but can digest if it undergoes a one-step mutation, hastens up its own mutation rate and survives. This phenomenon, directed mutation, was heralded by the popular Newsweek magazine as evidence for Lamarckian evolution. The biologist Dennis Todd and I (1997) explained the data of Cairns et al in terms of Lamarckism and biological group creativity.

Quantum Evo-Devo

In conventional biological thought, genes are the only hereditary material and therefore the only player in evolution; it is from chance genetic variations that nature selects in the evolutionary process. There is no room for any contribution to evolution from the acts of living. The prejudice of the central dogma of biology—information can only flow from the genes to proteins not the other way around—prohibits that.

And yet the fossil data shows repeated evidence that the same biological forms repeat again and again as evolution proceeds. These forms, in other words, are *remembered* in some way; some kind of Lamarckian mechanism prevails. So, there must be a side door through which development can enter the discussion of evolution; we must use the side door without challenging the central dogma which is supported by experimental data.

In quantum biology, the straightforward explanation of the repetition of biological forms over and over again is that the universal biological software consciousness bestows an organ with when the organ evolved consists not only of function but also of form as our name for the fields responsible for the software indicates.

In this way, the biological evolution of organs and their software also produces templets of form-making in one fell-swoop of quantum leaping; in other words, consciousness and the morpholiturgical fields not only provide the organs with functional software but also guide their forms during evolution. Like all quantum objects, once the software is manifested and collectively, for the entire species, it returns to the unconscious as nonlocal memory with increased probability for future use. Note: although software is the result of repeated use and involves some conditioning, the conditioning is far from 100%; some quantum behavior continues; in particular, nonlocality.

Organisms exist in homeostasis. When a crisis occurs, they respond as a group with intention-making for survival. When this intention resonates with the availability of a gestalt of possibilities for a biological expression such as an organ, the quantum leap takes place. With increased survivability, the organism survives its existential crisis as the crisis passes after eliminating all but a small percentage of the old life forms. After a brief period of adjustments (the manifestation phase of creativity), some of the new organisms settle down to a period of homeostasis of adaptation and continuous evolution. Often a flurry of biological creativity occurs in the aftermath of the crisis producing a period of rapid radiation.

Often a whole bunch of new organs or major modifications of the old take place. As for forms, consciousness uses old templets of forms whenever available. Each such period of rapid change will additionally produce a lot of new forms.

And then another geological catastrophe, again a lot of extinction, and another flurry of quantum leaping giving rise to a lot of modifications of old forms/functions and addition of new forms and functions. These new forms then will act as templets for forms to be used in future such periods of rapid evolution.

Avantgarde biologists (Carrol, 2005) have named this process appropriately *evo-devo*—evolution and morphogenesis take place hand in hand; *evolution is not only based on inherited genes but also on inherited forms.* However, without the quantum science, no mechanism of inheritance of form can be given; hence I rename this evolutionary process quantum evo-devo.

Old forms that appear early in evolution become potentialities for later use with higher and higher probability as they are reused again and again. Just like in any memory retrieval of human experience—recall enhances the probability of recall.

Vital evolution geared to survival more or less culminates with the evolution of the human being with the neo-cortical brain. The quantum in every form or organ of our physical body, the P-organ, comes correlated with a vital software—the V-organ—that guides the on-off of the gene activators that make proteins for the organ functioning (see fig. 5). I repeat: It is the movement of the V-organ in response to a stimulus that you experience as a feeling. Vital energy is the energy of movement (collapse or cessation of ongoing collapse) of the V-organ.

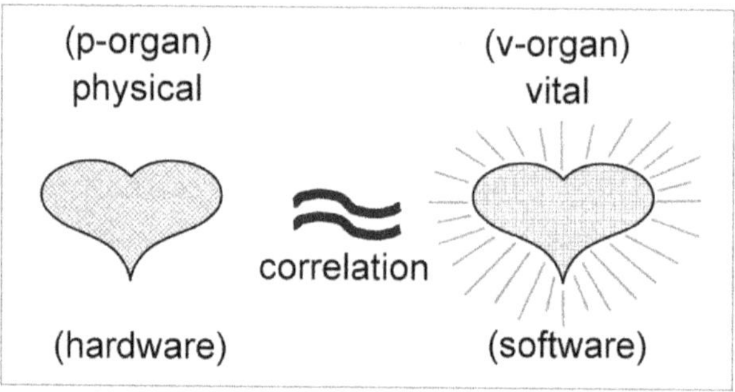

Fig. 5. Physical organs (P-organ hardware including physiology) come with a correlated vital counterpart (V-organ software)

Is there evo-devo in microevolution of Darwinian vintage driven by blind chance and survival necessity? Is there deviation from randomness? Recent findings show that even Darwinian evolution is not as random as was once thought (*https:/www.sciencedaily.com/releases/2024/01/240108125801.htm*); it seems to depend on history, conditioning. This is likely an effect of evo-devo.

Keep this important but necessary modification of Darwinism with you:

Darwin failed to incorporate Lamarck's idea of inherited characteristics in his theory, but his intuition about its necessity finally paid off. Yes, in quantum science the inheritance of acquired characteristics is incorporated in a scenario called quantum evo-devo: Darwinian evolution of genetic hardware coevolving with epigenetics—Lamarckian evolution of universal living developmental forms and functions gives us the organ hardware and physiology—P-organ. This is quantum evo-devo. This accounts for the repetition of certain biological forms and functions over and over again as evolution proceeds.

The P-organs are correlated with vital software—the blueprints for organ function via their parallel quantum gene activators (fig. 6).

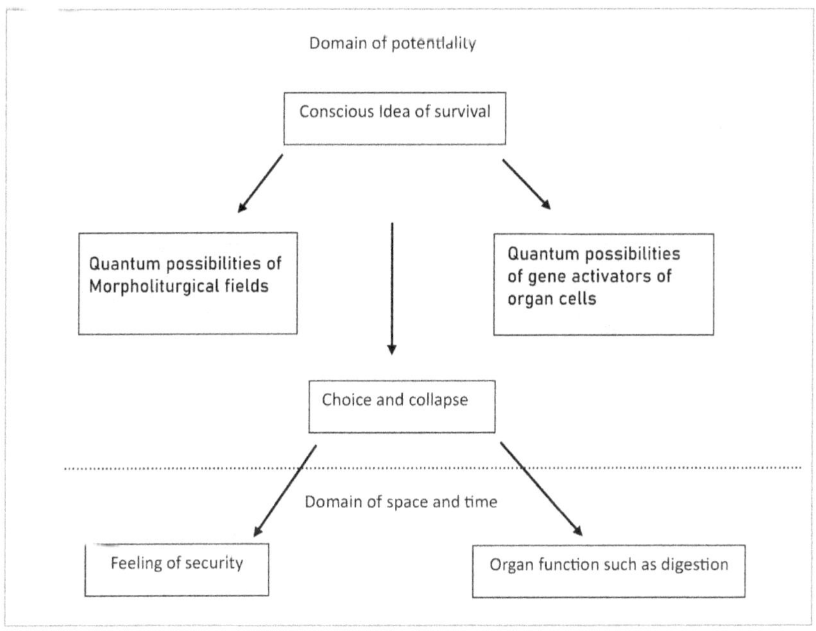

Fig. 6. Psychophysical parallelism: How feelings and organ functions are related

One final comment before passing. It is easy to make fun of Lamarckism and the idea of evolution via acquired characteristics. An unnamed author recently wrote a piece called *I Ain't a Joke* in *Illuminations Mirror*, July 19, 2023, issue that I found on the Internet:

My high school teacher laughed while describing the theory of Lamarck.

He used giraffes.

Giraffes can't stretch their necks and make them longer every generation.

It [Lamarckism] would mean mice whose tails are consistently cut in labs should have shorter and shorter tails over time. We can expect mice without tails after a few generations.

Jews could have had their male children born without the foreskin.

In Kenya, the Luo men would have had their boys born without their six lower front teeth.

The idea is laughable.

So my high school teacher laughed.

It was not my biology teacher. He had some decency. Maybe he knew the idea could be possible. Maybe not the lengthening of the necks, but another acquired trait. I never asked him.

Our theory of quantum evo-devo, I hope, has cleared up the situation for you. Acquired characteristics of organs that propagate are changes in the software of form and function, not changes in hardware.

The Quantum Science of Feelings: The Chakras

I have already introduced the concept of the vital body, a collection of universal vital software behind organ functions (physiology), the conglomerate of all the V-organs in us.

The vital software gives us a profound explanation of feeling: what we feel (the movement of V-organs associated with P-organs), how we feel (as the quantum energy of movement—vital energy— as consciousness collapses the software or stops collapsing it), and where in the body we feel.

Amazingly, the map of where we mainly feel during the course of our living was empirically discovered long ago in the form of the all-important chakras famous from Yoga psychology.

Biologists mostly concentrate their attention on the functional molecules of life—proteins and genes. Except for field biologists, they don't study life while the living goes on, their studies mostly are confined in the laboratory, outside of the arena of living. In this way, the biological mainstream misses much about life and living.

How do we study life without excluding living? Studying behavior as field biologists do, is a part of it to be sure, but that is only the outside component of living. There is also the inside component—feelings. Materialist biology, by ignoring the distinction of living and nonliving, also ignores the distinction of outer and inner in living, eventually ending up ignoring the inner component of living, that is, feelings, altogether.

If you watch the TV series *Star Trek,* you will remember how that robotic character Data struggles to incorporate emotions in his being. Alas! The writers of *Star Trek* eventually gave in to the simplistic notion of an emotion chip for Data. Most establishment biologists do think that emotions are brain phenomenon. There is debate about how much of emotion can be understood on the basis of neurology and molecules of emotions. Many biologists refer to that ultimate explanatory principle of Darwinian evolution to say something useful about emotion. But of course, the fossil data tell us nothing about what went on in the interior experience of living creatures.

Some biologists to their credit admit that feelings do not originate in the neocortex, that they are not computable. They even agree that feeling may occur before the cognition of thought during an emotion.

Sometimes, it is hypothesized that feeling and emotions are the territory of the neurochemistry of the limbic brain. To this end, the researcher Candace Pert's experiments on the "molecules of emotion" are noteworthy; read her book by that name. Certainly, these molecules are telling us something important—that there are molecular movements that we sense when we emote. That it is impossible to sense without simultaneously emoting. One of the conclusions when we consider sense perception is that is impossible to perceive without cognition (see chapter 1).

But is it all about the brain? Isn't it our experience that feelings also originate in the body? Otherwise, how would people of our culture perpetuate such notions as "butterflies in the stomach" or "heart-warming" when talking about their feelings?

In modern times, only William James, the father of American psychology, had the right idea that agrees with people's experience. According to James, feelings are associated with direct bodily changes when confronted with some stimulus.

Fortunately, yoga psychology of the old, feelings are recognized to be associated with the organs and their physiology and emotions are clearly seen as effects of feelings on the mind. According to yoga psychology, there are seven major centers of our body—the chakras—where we feel our important feelings. But through the centuries, although the idea and usefulness of the concept of the chakras have found much empirical validations from spiritual disciplines, not much theoretical understanding has come. Now, finally, with the ideas of morpholiturgical field and psychophysical parallelism, an explanation of the chakras, where our major feelings originate and why, can be given.

The Quantum Science of the Chakras

I have treated this subject in some details in many books; read the book *The Quantum Brain* for a recent review. Here I will be succinct.

The seven major chakras along the spine (fig. 7) are each located near major organs for our body's physiological functioning. Chakras are those points of our physical body where consciousness simultaneously collapses the function of one or more organs along with their vital software that guide the function. The feeling you experience at each chakra is of the vital energy movements of the vital software, V-organs at that chakra (see figure 8 later).

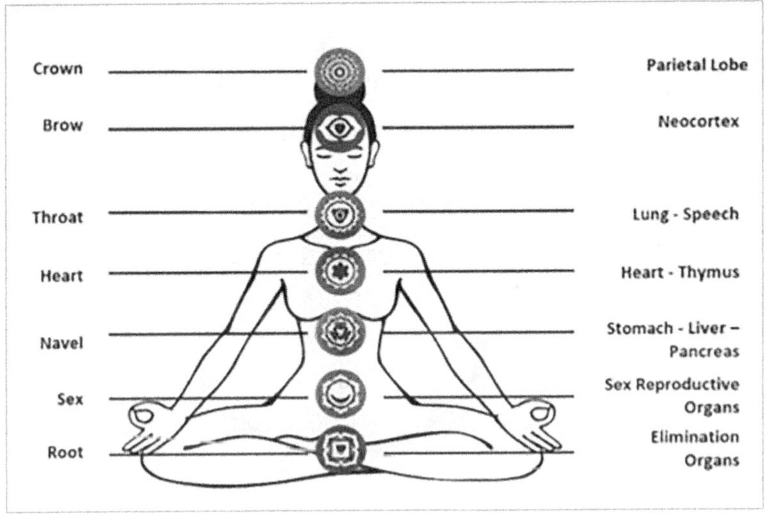

Fig. 7. The chakras along the spine and their associated organs

Here is a chakra-by-chakra description of the physiological functions, the corresponding physical organs, and the associated feelings due to the movement of their correlated vital software:

Root chakra: the physiological function of the organs at this chakra is waste elimination. The organs that express the vital software of the function are the kidneys, the bladder, and the large intestine (rectum and anus). The associated feelings are self-centered rootedness, survival-oriented competitiveness, and fear.

When you feel rooted or secure, the experience is a positive one; the organs are collapsing into actuality and functioning properly. Vital energy is being generated at the chakra. In the language of chakra psychologists, energy is moving *into* your root chakra. Notice the language used; it is not that they mean physical energy is moving into the physical organs at the root chakra. They mean vital energy.

On the other hand, when you experience fear, the chakra psychologist will say that energy is moving out of this chakra. In quantum science, we explain: the ongoing function of the organ(s) at the root chakra has been interrupted, temporarily suspended, in response to some stimulus.

With quantum understanding in mind, we will use the chakra psychologists' language because it better describes our experience. It is not that unfamiliar experientially, is it? People who are fearful by nature, cannot take risks, they also chronically suffer from constipation, anal retentivity. It is because their brow chakra of giving meaning is stuck at the root chakra, at the anus.

Even closer to home, if you see a wild animal in the forest, you will really experience fear as a feeling of energy leaving your body through the anus. In quantum lingo we would say, the physical organ function has stopped collapsing momentarily suspending the function. Often prolonged exposure to fierce stimuli of fear will produce elimination organ disfunction; people are found to urinate even defecate in their clothes.

Sex or sacral chakra: the biological function is reproduction, the vital movement of feeling is associated with the reproductive organs—uterus, ovaries, prostate, testes, penis, vagina, etc. The feelings are sexuality and amour when in response to a stimulus, vital energy moves into the chakra in an ongoing way—organs functioning properly; and unfulfilled lust when energy keeps moving out and the organs are not smoothly functioning.

Jives with your experience, does it not?

Navel chakra: the physiological function is maintenance and the organs are the stomach, small intestine, liver, gall bladder, and the pancreas. When energy moves into this chakra, you feel self-assured, when energy moves out, you feel unworthiness, unsure of yourself—butterflies in your stomach. Rings a bell?

Here is something interesting. If energy moves into the navel from the sex chakra (unfulfilled lust or desire), we feel anger. Remember: in quantum language, this means, we are distracted from using the sex chakra and feel the energy at the navel chakra instead. Unconscious unfulfilled feelings of sexuality lead to anger.

Why do many Americans watch so much sex and violence on the movie screen or television? Americans today live a very head-centered life; vital energy is experienced always up there at the brow chakra right behind which we have the thinking brain—the prefrontal cortex. Could it be that sex and violence help to bring down the vital energy to the bottom three chakras? People can feel rooted that way but this tactic also brings horniness and eventually anger and even violence, if the sex goes unfulfilled.

Heart chakra: the physiological function is self-other distinction (between what is perceived as *me* and what is perceived as *not me*) and defense against the "other." The organs located at the chakra are the heart and most importantly the thymus gland. The thymus gland is the crucial component of the immune system whose job is to distinguish between what belongs to my body (me) and what do not (not me) and get rid of intruders. Feelings of defensiveness are the result when the immune system functions well. When energy leaves the chakra (that is when the thymus gland function is momentarily suspended), the feeling is vulnerability.

Chakra Awakening at the Heart and the Navel

The heart is traditionally associated with love; is it just folk lore then just as the materialists claim? No indeed.

For people of base-level condition though, the heart is just a blood pump, a function that is not connected with any of our important feelings. Yet some people undoubtedly feel *romance in their heart* when they meet a suitable person usually of the opposite sex. What mechanism is responsible for falling in love in this way?

My collaborator Dr. Valentina Onisor and I have discovered the secret (read our book *Quantum Integrative Medicine*). With the ongoing presence of a suitable person of the opposite sex, an intention quickly precipitates an intuition for developing intimacy that grows in thus affected people; in this way, when they interact with their intended lover, they allow vulnerability and suspend the thymus gland function momentarily in an ongoing fashion, whenever they interact. Following up their intuition via the creative process leads them to a quantum leap of the heart. New morpholiturgical fields come into play; the quantum gene activators of the individual cells of the heart start acting in coherent quantum movement; in other words, what was ordinarily quantum movement at only a few cells and incoherent at that, now there is quantum movement at the macrolevel of the organ itself. Such macroquantum coherence has been experimentally observed by researcher at the Heart Math Institute in California.

The macroquantum movement of the gene activators activate new genes and the heart makes new proteins producing a higher physiology that include being sweet to another person—love.

You feel the sweetness of romance when your consciousness expands to include the person (usually of the opposite sex), you cease to distinguish between you and him (who is no longer perceived as not you), because your immune system goes through momentary suspensions when you two interact. If on the other hand, you are in romantic love with someone, and he is paying attention to another person of the opposite sex, you feel the energy leaving the heart chakra, causing you the feelings of jealousy and hurt. You feel hurt because your heart has gone back to being a blood pump.

Now this one should ring a bell in everyone because who doesn't remember being a teenager and experiencing romance as heartthrobs? Or is there anyone who has not experienced jealousy? When we get older, the romantic feelings abate a little and are experienced as tingles or just warmth.

When the expansion of who you are happens, it is possible for you to give anything to your partner, even your life. Only thing is, the phenomenon of romantic love gets additional boost from the emission of neurotransmitters, principally dopamine. When the dopamine production trickles down to a minimum, your attention will stray. The heart may still feel warm; but the molecular component of pleasure is gone. This is why for the fictional movie character, James Bond the romance lasts as little as six weeks. Yet we like to watch his escapades because during those six weeks, he seems to be able to repeatedly endanger his life for his woman.

We have discovered one more thing. There is a little brain—a big bundle of neurons at the heart chakra. In this way, when the heart becomes macroquantum it can process stimuli (arriving nonlocally from other living beings) with both cognition and memory-making capacity and thus tangled hierarchically. In other words, consciousness now can identify with the entire heart chakra at the organ macrolevel.

Similar things happen at the navel chakra as well where also there is a little brain. The organ that becomes quantum at the navel is the pancreas. The role of the thymus gland is now played by the stomach. When the stomach is starved (fasting), your intention to love yourself may be intuited. If you follow this with creative do-be-do-be-do, the pancreas makes a quantum leap; its quantum gene activators start moving coherently and activating new physiology that enables you the new feeling of self-esteem.

When excess energy moves into the heart from the lower chakras, from the sex chakra after sex and the navel chakra after a good meal, you feel giving, you feel vulnerable to your own generosity. Men specially, and many men suppress heart energy because they don't want to become vulnerable to others, even to their romantic partners.

Please remember this in passing:

> Many people (men mostly) do not experience romantic love much at all; romantic love is by no means universal.

> Similarly, many people (women mostly) do not want to love themselves; so conditioned they become through sociocultural conditioning and perhaps childhood trauma.

Vital creativity is required to practice vulnerability (the result of temporarily suspending the thymus gland) or desire for self-love (to momentarily deprive yourself from the pleasure of eating) and then engage do-be-do-be-do with the archetype of love and then only a quantum leap takes place, the heart or the pancreas becomes all-over quantum, and new morpholiturgical fields come into play giving us the feeling of romance. The organs also function with "higher" physiology.

Throat chakra: The biological function is self-expression and the organs that represent it are the lungs, throat, and speech organs, the organs for hearing, and also the thyroid gland. When energy moves into this chakra, when the chakra becomes active, we feel the exultation for our freedom of expression. We feel the opposite, frustration, when energy moves out. Our throat dries up when we are unable to express ourselves.

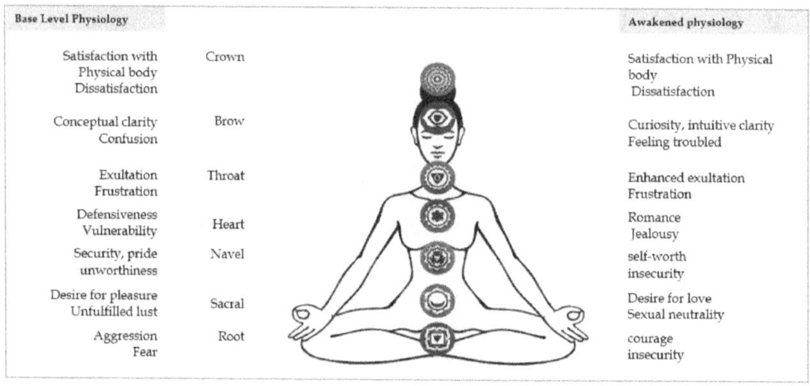

Figure 8. Chakras and how their associated feelings change when the brow, heart, and navel selves owoken. Feelings for the base-level physiology are shown on the left; feelings for the awakened physiology are shown on the right: For each chakra, the top feeling signifies positive feeling when the chakra-organs are functioning well; the bottom feeling indicates negative feeling when the chakra-organ function is disrupted.

Brow chakra (also called the third eye when it "opens", that is, when the organ engages new software of higher functioning): the biological function is thinking and the associated organs are the neocortical organs (of mainly the prefrontal cortex). When the appropriate organ of the brow chakra—the prefrontal cortex—takes a quantum leap (or as per chakra psychology, "fully opens), the physiology changes to new functions of intuition and creative insights; we have something new to cognize—the archetypes—so the chakra is now also called the third eye—the eye of intuition.

Many young people feel warmth at the chakra when the third eye opens for the first time. That is how in India, the tradition for girls to put a *bindi* (covering) on the brow chakra (fig. 9). Even when the chakra remains only part open, when you concentrate on rational thinking, you will notice you brow chakra muscles will tend to contract indicating that your vital energies are high there; you may feel passionate about what you are thinking and you may gain some clarity.

Figure 9.

Conversely, when energy moves out of the brow chakra, we feel confused. I make the point in this book that right now our evolution in consciousness is blocked. Naturally, we as the entire humanity are confused. You will notice if you pay attention, today there is a general lack of passion for most people of our society when they think anything other than of themselves.

Crown chakra: the physiological function is body-image. The organ is the parietal lobe of the cortex where all the organs of the physical body is mapped by the somatosensory input and responses. When energy keeps moving into the chakra, we feel satisfaction with how our body looks; if energy moves out, we feel distraught.

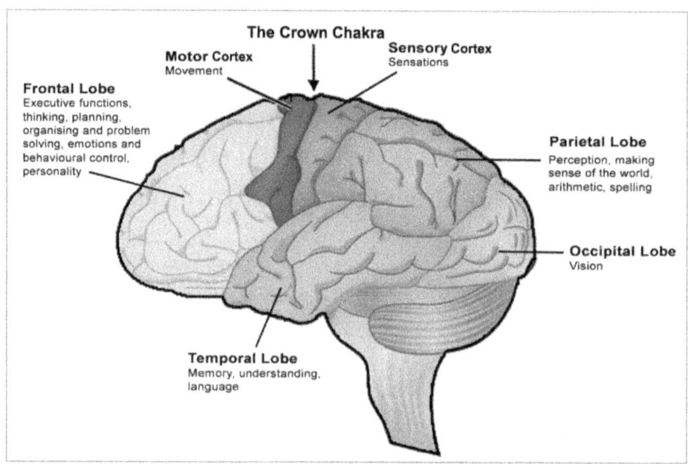

Fig. 10. The Crown chakra

There is also the pineal gland at this chakra and that likely comes into quantum play with quantum leap (complete opening of the chakra) when the parietal lobe function of body-attachment is momentarily suspended.

Even with this cursory intro to the chakras, I hope you can see that the materialists have it all reversed. They think that we feel emotions only in the brain, that is, emotions are brain epiphenomena. And then it comes to the body through the nervous system and the so-called molecules of emotion. But actually, it is also the other way around. We can receive vital energy from other living beings nonlocally and feel pure feelings at the chakras. Yes, the control goes to the mid-brain for integration through the nervous systems and the molecules of emotion and eventually, the neocortex gets into the game when mind gives meaning to the feelings creating the experience of emotions. Keep these ideas with you along with the details of the feelings at the various chakras.

The Quantum Evo-devo of Animal Consciousness

Let's go back to evolution, quantum evo-devo. According to quantum biology, the evolution of the human being is the culmination of vital evo-devo, basically synonymous with the culmination of the evolution of the rest of the entire animal kingdom. An important question to ask is this: how has the animal consciousness evolved empirically, the details? What does the data indicate? All life begins as a single cell species and then evolves to multiple cell organisms which then split into three kingdoms—the fungi, the plants, and the animals. Animals develop organs. Do animals experience their consciousness at the chakra organ level as soon as organs appear in the evolutionary scene?

When I researched the subject, my initial assumption was that the animal evolution of conscious awareness should proceed chakra by chakra. Animals with a developed digestive and reproductive systems should have conscious awareness via feeling at the first three chakras. Then animals should progress and have consciousness awareness in all four chakras in the body. And so forth.

An examination of the data (as much as possible for a physicist; remember I am not a biology expert) was disappointing. Animals until they reach the level of domesticated mammals exhibit only machine-like behavior; whatever little conscious awareness of feeling they have is readily explained by consciousness at the society-of-cells level—multiple cells held together by both local and nonlocal interactions. Why don't animals have chakra consciousness? Are animals then basically mechanical in behavior as Christianity has posited?

Only recently did I become aware that even us humans have self-identity at the macrolevel in the torso only at the navel and the heart chakras. This realization followed my discovery of new data that little brains in the form of big bundles of nerves are found at only those two chakras.

Let me repeat. The conscious experience of subject-object awareness the way we do have at the brow chakra, the site of the neo-cortex, requires a tangled hierarchy between the cortical perception and memory apparatuses. A memory apparatus consists of a bundle of nerves for making memory circuits. Since such bundles of nerves exist in the torso only at the navel and the heart, tangled hierarchies and self-identity at the macro-organ level can arise only at the navel and the heart.

This does not negate the chakra theory. Vital energies at the other chakras will collapse at the community-of-cells level and the information will go to the midbrain for further processing.

Animals evolve the nervous system pretty late in the game and a bundle of nerves big enough to form a memory apparatus does not seem to appear before the advent of the mammalian brain—the midbrain or limbic brain.

Ok, this explains the field data—the lack of organ level animal consciousness until the mammals come into the scene.

If we posit that there is a tangled hierarchy at the mammalian brain and self-identity at the macrolevel does arise there this would be a prediction that quantum science enables us to make. A chakra with macrolevel self-identity at the mid-brain.

Evidence for the Mid-brain chakra

The problem with the proposition of a chakra with self-identity at the mid-brain is that we humans, who should inherit it from our mammal ancestry, do not seem to experience it. So much so that neuroscientists of the materialist ilk call the memory located in the mid-brain "unconscious" memory. Chakra psychologists of the old underemphasize this mid-brain chakra as well perhaps because they, too, agreed with the assessment of today's neuroscientists.

But quantum biology enables us to think otherwise! It is true that for most people, the cortical-self overshadows the midbrain self so much so that the latter seem to be unconscious most of the time. However, there is exception.

There are people in whom the midbrain self is quite active. Obviously, only when the cortical self is weak, it would be possible to experience the mid-brain self equally well. The behavior of such people would be perceived as deviant from "normal."; sometimes it is even labelled as a personality disorder. Could this schism give rise to pathology?

Indeed so. I have formulated a theory of schizophrenia as a development of a real schizo—divide—between the cortical self and the midbrain self (Goswami. 2024; also, read my upcoming book with Valentina Onisor, *Quantum Psychology and Mental Health*). Anyway, if this theory is verified in the future, that would prove the veracity of the idea of a mid-brain chakra with self-identity.

The idea of mid-brain chakra of self-identity could be big as far as animal consciousness is concerned. Considering that many biologists are still stuck with the idea that animals—even monkeys—do not have self-awareness. There is simple test. Put paint on a monkey's chest and show the monkey its image reflected in a mirror placed before it. A monkey cannot comprehend that the red mark is on its own body clearly demonstrating that it does not the capacity to identify its mirror image with its own physical body.

How do we then justify the claim that mammals can cognize, they have a self? You have to realize that the mammalian experience of the self, experience of cognition is based upon the experience of feeling at the midbrain chakra. Mammals, with self-identity located in the mid-brain can cognize only with feeling.

How to explain the mirror experiment mentioned above. Cognizing that the red mark is on its own chest requires a monkey to have thinking consciousness. A feeling consciousness cannot discern such details. Figuring out that the mirror is reflecting your body takes thinking, a capacity that these animals do not normally have. By the way, it takes human babies to pick up such body awareness as demanded by the mirror experiment a full year of development. Before that age, human children also cognize mainly with feelings. This confirms the theory proposed here.

To be sure, recent research concedes this much: after rigorous training at least, some monkeys can become self-aware. The training is needed to actualize the monkey's thinking potentiality.

There is one big consequence of the postulate of a mid-brain chakra in mammals. The vital energy actualized in the lower chakras of the body can now be received there through the neuronal connection. In this way, the brain takes over the functions of the sexual, digestive, and defense systems; this leads to the instinctual behavioral responses called the 4 F's—flight, fight, feeding, and f-king which even us humans have inherited.

Note this: the midbrain takes over the heart chakra immune function as well. Psychoneuroimmunology, the science of such takeover in humans and rats is now well developed.

Rats belong to that special category of domesticated animals. Does this takeover take place in all mammals, even wild ones?

A big question of animal consciousness has always been this: why can only some mammals be domesticated? A wild mammal refuses all efforts at domestication. I think a tentative answer to this question can now be given subject to further research. Only the domesticated animals have enough of a bundle of nerves at the heart chakra so that they can be trained to quantumize the midbrain chakra and awaken higher physiology and altruism at the midbrain chakra like us humans. The wild mammals—tigers and lions—do not, cannot take such a quantum leap.

Indeed, if this speculation is correct, it would explain all the affinity and love we experience with pet dogs, even cats.

As pet owners know very well that their pets—dogs and cats especially but even horses and elephants, respond to love with a self-identity that is unmistakable.

It is a fact that dogs are capable of enormous loyalty and love for their masters. Conditioned love of course, but impressive nonetheless. And they do feel; Rupert Sheldrake (1999) made some great videos demonstrating this. A dog waits on a sofa through the day, but when his master gets up from her desk to come home, the dog goes to the window. How does the dog know? Via nonlocally communicated feelings at its heart chakra.

Consciousness in primates, dolphins, and Human Children

How about animals with a neocortex such as primates and dolphins? Empirical research shows that chimpanzees can even be trained to think mental meaning. A lady chimpanzee was taught sign language for a bird and also for water. When she saw a duck, she herself joyfully constructed the sign language for water-bird by combining those of bird and water.

The neuroscientist John Lilly who did a lot of research with the cognitive abilities of dolphins told me that dolphins are quite capable of being trained to acquire enough language comprehension and thinking to carry on a conversation with him the human, but he also acknowledged the problems with such research. Dolphins in confinement get bored with everything; all they care about is sex.

Let's talk about babies, human babies. Empirical research also shows that most babies acquire rudimentary thinking self-awareness by age 1. By age 5, they have what can be called a mental ego with enough repertoire to get by most needed tasks. And of course, a child's mental ego continues to develop becoming an adult ego by about age 18-21.

The ego evolves as the memory repertoire, especially at the hippocampus in the midbrain, becomes increasingly larger, this much is clear. Initially, when the ego is weak, the child's consciousness does not identify with the neo-cortical memory completely. The identity becomes more and more complete as the memory apparatuses (mainly the hippocampus) can sequence events fully; what happens then is that the forgetfulness of the quantum self becomes more and more compulsory. I have a read a story of a three-year old elder brother talking to his two-year old sister to enjoy her conversations with "God," as long as she can, because "I am already forgetting how to do it." By age 5, this forgetfulness sets in for all practical purpose and the ego-me starts to become supreme, the head honcho of growing programs of personality. No wonder that even in the spiritual culture of India they regard babies as God only until they reach age 5.

Keep these important new ideas with you particularly if you like animals:

> *Only mammals among all the animals are conscious at the organ level; this quantum science conclusion is based on the prediction that there is self-identity at a chakra in the midbrain, the mammalian brain (Goswami and Onison, 2022). However, remember that among animals, only mammals have midbrain, and even they can normally cognize only with feelings.*

> *Animals before the mammalian brain appeared on the scene, are a society of multicellular individuals in spite of having organs. At a local level, they are able to act individually. Additionally, they have a nonlocal group consciousness for the entire group.*

Brains and Selves in the Body

Let's elaborate on the important subject of the selves in the body in humans mentioned above. As I said earlier, I had been worrying about that ever since I discovered the quantum theory of feelings at the chakras.

A short while ago, my collaborator Valentina and I noticed that the researchers at the Heart math Institute have pointed out that there is a huge bundle of nerves that connects the heart chakra to the brain cortex. They, too, have been wondering about that.

For the quantum way of thinking, the presence of a bundle of nerves, a sort of a "little brain" at the heart chakra means memory-making capacity at the chakra. We have been noting that the heart by virtue of our developing the feeling of vulnerability that comes with immune system's momentary suspension develop the ability to cognize love; add the capacity of making memory to the cognitive capacity, what do you get? Both cognition and memory making a tangled hierarchy, right? And that means only one thing: there is a self of the heart after all.

And guess what? There is a similar large bundle of nerves, a little brain, also at our navel chakra; so, by the same argument, there must be a self of the navel as well. Put together what I concluded above. There is also a naval self.

First of all, most people in modern times don't feel energy in the body much at all; so, they miss the experience of pure feeling for this reason. When you acquire sensitivity to feelings in the body, that is one good step. So, males begin to feel their self in the navel; all male dominated culture identify a "self of the body" at the navel. For example, in Japan, they call it *hara* and suicide they call *hara-kiri*; killing yourself is killing your *hara-self*.

Second, women of course usually have some awareness of the self of the heart; that's why they swear by it often to men's consternation. When I got married a second time, whenever I said "I love you," to my woman, she would complain "but you are not saying it from your heart" and I would be befuddled. What is she talking about?

Third, spiritual traditions often talk about male-female integration. What do they mean? In the broad sense, we interpret that as the integration of reason and feeling (emotions). But in truth, integrating the male and the female of us is not so much about that, it is more about integrating the male body-ego at the navel and the female-body ego at the heart. If we don't integrate, we become dependent on the other; males depend on women to take care of their relationship which requires other love; females depend on men for worldly matters, such as earning a living (which requires self-esteem).

Take these important discoveries of quantum science with you:

There are two selves in the body, one at the navel, and at the other at the heart; when men talk about gut feelings and women tell you about loving from the heart, take that seriously; they are scientific ideas.

The male-female difference that spiritual traditions of the old emphasize arises from the males emphasizing the navel-self experience and females the experience of the self at the heart. It behooves us to integrate this dichotomy for our optimum functioning.

PART 3

Evo-Devo of the Mental: the Ascent of Humanity

CHAPTER 8

The Quantum Evo-devo of the Mental

The evo-devo of the mental is how the ascent of humanity takes place with consciousness using the mind to progressively give meaning to the four increasingly subjective human experiences—sensing, feeling, thinking, and intuiting. In this way of reckoning, the era of the intuitive mind is in our future. But how do we go there from where we are? mind must turn to the final challenge: giving meaning to the self-experience. If the 15% of humanity that is already exploring the potentialities of the intuitive mind, go for community building, there is hope for quick change. In this final part of the book, I will try to construct a scenario for the advent of this glorious future of humanity.

If the mind were an epiphenomenon of Darwinian evolution and of the brain, then any evolutionary changes in the way the mind operates could result only from evolutionary changes in the brain. Also, such changes could not be progressive, could only be haphazard. But this is not what is empirically observed.

Sri Aurobindo and later Ken Wilber theorized that how we use our mind—mental software—evolves through four stages: the physical mind—mind gives meaning to the physical and embodies it—producing the sensory-mental software; the vital mind—mind gives meaning to the vital and embodies it as emotional software; the rational mind—mind gives meaning to the mental itself, this gives us mental software with which we do abstract rational thinking; finally, the intuitive mind—mind gives meaning to the supramental archetypes that we intuit producing soul software.

What does our quantum evo-devo model say about the stages through which the evolution of the meaning giving facility of the mind must proceed for the entire human species? In quantum theory of evolution, in contrast to animals, we cognize mainly through the mind; we start with the three possible modus of cognitive experience with our inherited physiology—sensing-thinking responsible for sensory-mental software, sensing-feeling-thinking responsible for emotional software, sensing-thinking-abstract thinking responsible for mental software, and end with intuition-thinking-feeling soul software of higher physiology responsible for soul level of being.

Although the brain and perhaps even the body hardware has evolved in important ways during these stages, it is the development of the mental (and the vital) software that takes the center stage, so here we will emphasize the devo part of evo-devo. This gives us evolutionary stages of the mind as more or less in the same chronology as in the original conceptualization of Sri Aurobindo and Ken Wilber.

Indeed, in the developmental history of humanity as anthropologists have put together, there have been so far three distinct stages that can be recognized as the era of the physical mind, the vital mind, and the rational mind, in that order. They were brought about principally via technology developed by human beings using creativity.

The Era of the Physical Mind

What is a physical mind? Consciousness uses the mind to put meaning to what the brain takes in as physical stimuli giving a sensory-mental response. The mapping of the mental meanings produces developmental mental software in the physical brain for which the physical brain memory (part of the living hardware) acts only as a trigger for the correlated mental memory (mental software) to play out. It is this mental memory maps that can be triggered again and again by the same or similar physical stimuli that we call the physical mind.

The earliest anthropological data of humans suggests that humans were hunters and gatherers initially; the males were the hunter and women, the gatherer; although this classification may not have been rigid. In any case, the physical stimuli occupied the minds of people of such a society the most; it helped survival greatly for both hunters and gatherers.

However, the preoccupation of pregnant and young mothers with unavoidable emotions among other factors rendered women in general into second-class status and hence patriarchy developed. The males were the architects of the worldview of this genre; the male "priests" had the job of explaining the physical world around them. Whatever they could not explain, such as natural forces of wind, thunder and lightning, etc. they attributed to nature gods, naturally all male.

Sociologists have designated the worldview of people of this era as magical. This is clearly justified.

For some people to this day, even after the development of the adult ego, the dominance of the physical mind continues. It is operationally okay to get by with the affairs of the world with the physical mind alone, but such a person does not access the rich, inner life available today and is susceptible for magical thinking that includes what today we call conspiracy theory.

I have spoken of theoretical scientists before; they are sold on the rational mind. But the empiricists among scientists tend to have an overdose of the physical mind as well; for them their rationality comes second. A person with such a mind is capable of building a logical structure no doubt and they do that only when he or she can model it from the sensory world of physical objects. Whichever cannot be thus modeled this way they consider magical; they call it supernatural, not to be accepted as part of nature or reality. Hence their refusal to see eye to eye the quantum worldview and the subtle human experiences such as the archetypal that the quantum view emphasizes. Sad indeed, but how predictable.

It can be and is actually worse with the intellectual ones for whom the rational mind dominates with the physical mind as a close second. People of this kind of mind are resistant to the acknowledgement of not only the archetypal intuitions but also feelings because they cannot make neither a rational nor a physical model of either. These people today live in a prison of cognitive dissonance of their own making with constricted consciousness wearing the straitjacket of scientific materialism which they think explains everything, because, "what else can it be?"

But here, women have an option. If women choose motherhood, even after growing up in a rational/physical culture, they cannot ignore nurturing children that requires attending to feelings, feelings in the body chakras, sometimes in way independent of the thinking brain/mind. Thus, partially liberated from the physical/rational mind, if they choose, they can help liberate the entire culture. This is what is happening in today's women in some of the developing economies such as Brazil, India, and Africa; they are taking the lead in this department. Women are taking the lead in bringing emotions back in America and Europe, too, but their ranks are divided.

The Era of the Vital Mind

The vital mind consists of mental meanings of feelings that arise in response to physical and vital stimuli. This is the era when instinctual feelings of animal ancestry were experienced as negative emotions and led to the production of the negative emotional brain circuits of our midbrain.

There is no contradiction between the ways of the physical and the vital mind. Also, for a vital mind, the physical mind continues to be useful and is therefore integrated in.

One characteristic of the vital mind is to look at the world as a dichotomy of likes and dislikes, positive and negative. Among other things, this is a precursor to the good-evil dichotomy of the mind that came later in this era.

The vital mind was clearly the way of the human mind in what anthropologists call the horticultural era or the era of the family-run garden agriculture-—a farming culture using simple machinery like hoes and spades. This is one era when both men and women worked at the same occupation—farming.

Picture men and women working together all day with all those instinctual feelings of animal ancestry that arose between them as they faced stimuli together and their minds gave meaning to those feelings creating emotions somewhat differently. As a result, their vital minds were lost in the melodrama of instinctual emotions that they mostly experienced negatively. In this way, the vital mind of this era thrived in sensuality, pain and inflicting pain, and pleasure. But that was not all.

Since feelings dominated both men and women but women had a clear lead in terms of familiarity with feelings, especially positive feelings, this has been the only age (before the present late stage of the rational era) when women had the same social status as men and when matriarchy perhaps was as much or even more prevalent as patriarchy. Since nonlocality of the vital pervaded the social structure, the individual egos were nonlocally connected when the occasion demanded, and the families and societies did not have much simple hierarchical structure.

Indeed, anthropologists tell us that whereas the previous era of the physical mind had only male gods, in this era one finds both male and female gods being worshipped reflecting the equal status of men and women. And also, the gods must have transformed from nature gods to archetypal gods; clearly, the archetypes were discovered, explored, and embodied leading to a few positive emotional brain circuits. Also, clearly, women were allowed to interpret the archetypes and their interpretation was valued. It is quite appropriate that this era is regarded as the golden age by ecofeminists. The Indian wisdom tradition did even better: they called this era *satya yuga*, the era when truth of the archetypes prevailed. This is virtually the same concept as the garden of Eden in Christian mythology.

This reminds me of a joke. A little girl was curious, "Mother, how did the human race come about?"

Mother answered, "God created the human race dear—Adam and Eve who lived in the garden of Eden. Then they ate the apple, were banished from the garden, had children, and all the humans you see today is the result."

Now the little girl went to her father and asked the same question. Father said, "Well, the human race came from monkeys."

Now the child was confused and went back to the mother. Mother explained, "I was talking about my side of the human family, and your father was talking about his side."

Joking aside, clearly, people of this era made a breakthrough in their attempts to explore the archetypes. And this era could very well be regarded as the beginning of the human race, when humans truly became different from animals .

But the vital mind is resistant to the rational mind, so it cannot engage the intuitive facility fully which would require acting synchronously with the rational mind to sort out feelings with rationality according to the archetypal and meaning context in which feelings arise. So unfortunately for us, the people of this era made the archetypal representations with the two valued-ness of likes and dislikes. Naturally the archetypal representations became dichotomies. The most famous of these dichotomies and the most pervading is the good-evil split.

Later in the rational era when emotions were repressed, these archetypal dualities were relegated to the unconscious along with the instinctual negative emotions as well as the few positive emotions mentioned above. The repression was so deep it remained unnoticed until Jung discovered these representations of the archetypes, they are now called Jungian archetypes and the part of the unconscious they reside in is now called the collective unconscious. Psychologists soon recognized that gods and goddesses of our mythology are also archetypal representations; they too reside in the collective unconscious.

Also, naturally, the worldview here changed from magical to mythical centered around the myths about the gods and goddesses and the human struggles with the good-evil dichotomy.

The Era of the Rational Mind

When we use the mind at purely the mental level without reference to the physical or the vital, when we learn to analyze the meaning of meaning, we are in the rational mind. At this stage of the rational mind, we are capable of abstract thinking and reasoning.

The logical structures of the rational mind do not have to depend on mechanical models, although they can. A person of rational mind is capable of looking at the meaning structure of his or her constructs, and also of engaging creativity and taking an occasional quantum leap of creativity to discover a new meaning.

The rational mind is clearly hierarchical; the head of the hierarchy is the personal ego that uses it to sort out its experiences and programs.

Anthropologists tell us that the evolution of the rational mind began with the change of technology from garden agriculture to large scale agriculture made possible by the invention of the plough. It must have been premature; the job of the vital mind was not finished. This is why to this day in most people the vital mind is not integrated with the rational. This is one reason that rationalists have such a hard time giving proper role to emotions in their lives.

The Biblical story of the "fall" is about the abrupt change from the vital mind to the rational mind. In the vital mind both Adam and Eve are in paradise. With hardly any ego development, they are one with nonlocal consciousness—they walk with God and talk with God. But all that changes when they eat the fruit of the tree of knowledge and rational mind begins.

I have heard. If you read the actual story carefully, the Aramaic version, you will find that it is Adam who was banished from paradise, not Eve. This is important in view of the fact that women were mostly denied meaning processing in the era of the rational mind until recently.

In the developmental history of the people of the rational mind, typically as children, people take many creative leaps of discovery (fundamental creativity) into the supramental to discover the contexts of (abstract) thinking. But as an adult ego, a homeostasis sets in, and such a mind identity becomes complacent with the learned repertoire of logical structures or belief systems. At best people of fixed belief systems are capable only of situational creativity.

Consequently, a rational mind displays the tendency of disinterest in the supramental origins of its learned repertoire of emotional contexts and in this way, the simple unconscious dichotomies—good and evil, beautiful and ugly, true and false, love and hate—and the negative emotional brain circuits dominate the behavior of supposedly rational people.

Feelings are not rational and thus the vital mind is not compatible with the rational mind. This plus the complacency of the people of the rational mind prevents them from integrating the vital mind which they now see as a nuisance. Women have a clear disadvantage in a rational society because of the importance of emotion in their lives, especially if they choose motherhood.

In the agrarian era men became the producers and consumers of meaning and once again as in the era of the physical mind, women were delegated to secondary citizenship of being primarily reproducers and child-caregiver. As such, they were not even allowed the privilege of meaning processing. Even among men, only the dominant men of society got to process meaning in a large way. This is the era that created traders and businesspeople and the great kings and warriors engaging in a primitive self-centered way with archetypes of abundance and power respectively. Soon all this gave rise to a clearly class-based society.

The affluence of the elite classes of society in the agricultural era allowed a new class of creative people exploring the more transpersonal archetypes such as goodness, justice, and wholeness—we can call them brahmins but the concept need not be restricted to Hindus. Unencumbered by the requirements of making a living, these creative people engaged in many hero's journeys of consciousness that created great human civilizations on the one hand and gave us a glimpse of the human potential on the other. This happened not just in India, but in many parts of the world: Egypt, Greece, China, Middle East, Italy, etc.

When the agricultural era was replaced by the industrial era with the advent of modern science, the era of the rational mind continued with new vigor except that the hierarchies gradually gave way to a more equitable sharing of meaning processing through the creation of such great social institutions of human thought as democracy, capitalism, and liberal (dogma-free) education. This began the "age of enlightenment" and a rational/scientific-materialist predict-and-control worldview began its journey to become one of the two competing worldviews that we see today.

Finally, with the advent of technological societies, when women forced their way into meaning processing, thanks to the technology of birth control and women's liberation movement and perhaps other factors as well, the rational worldview in the West faced competition. Women demand attention on women's values—positive emotions like love. Spirituality and values in general got a boost as well because of a large influx of wisdom teachers from the East.

In the West, science has a long history of feud with religion. The scientific elite embraced scientific materialism to deal what they thought would be death blow to religion. Women's lib was coopted in the rational culture by offering women better entry into the professional arena.

However, the inevitable backlash of people of religion and opportunistic politics created the current worldview polarization and civilization itself stalled creating crisis. An evolutionary pressure from the future, from the next stage of mental evolution—the intuitive mind began to show as considerable scientific interest turned toward inquiries about the nature of consciousness.

How the Archetypal Exploration was Compromised in the Rational era: the Varna System in India and Elsewhere

Archetypes have been discovered even before the advent of the rational era. However, for new quantum monads, which archetype attract them the most? Abundance, and that limited to material abundance—survival need. Then after a while, the surviving quantum monads turn to the archetype of power, but still pursue it only in the service of me.

Even today, human history as constructed in the West, is a story of the me-centered escapades of kings and the likes. Shows you how many of us are still lost in the pursuit of survival needs in the West.

In the East, especially in India, the spiritual history is available in some of the great literature of ancient India. I have previously mentioned Ramayana, Mahabharata, the puranas. There we find the developmental history of higher more spiritually developed people. The people of the archetype of abundance looking for material wealth developed first; they were called *Vaishya*. The people who seek power in the pursuit of personal security were next to develop; they were called *kshatriya*. And then only we find people mentioned who pursued more transpersonal archetypes such as goodness, truth, and justice; they were called Brahmins.

This is the basis of the *barna* system in India; people came to be classified according to their quality of their creative ability (gunas) and the propensities they develop (karma).

If we look with this added insight at the progress of Western culture, there too, we can see "brahmins" in the early Greek and

Roman civilizations in the form of philosophers. Later, there are not only philosophers but also people of morality and ethics who could be seen as brahmins. And of course, there have been scientists and artists always and everywhere who pursue the archetypes of truth and beauty respectively and serve higher needs. There also are brahmins.

In general, people who explore archetypes for higher need must be recognized as brahmins; professional people who explore archetypes of abundance and power at the service of lower need of survival must be classified as Vaishyas and kshatriyas respectively; people who can only do service jobs must be recognized as *shudras*. In this way, all societies have the barna system

I must sadly acknowledge, however, that this barna system eventually degenerated into the hereditary system in spite of the democratic declaration "all people are created (with) equal (potential)." All societies so far have failed to institute equal opportunities to explore the human potential to all people. Hindus have been the worst; they institutionalized the hereditary "caste" system in the form of a dogma in Brahminism—the exoteric popular version of Hinduism.

The Way out of the Worldview Polarization

Coming back to the "brahmins" of the rational era, the inertia of the purely mental rational mind is overcome by some of the brahmins often beginning with intense curiosity in some field of knowledge. And an occasional foray of fundamental creativity in a narrow field begins.

In that narrow field of expertise, these brahmins of the rational mind usually with a good amount of reincarnational maturity under their belt, take quantum leaps to the supramental to tackle primarily problems of "outer" arena; in other words, they participate in outer creativity still fascinated with the outer world and manifestations in it. Even if these people get an insight about the nature of love, they would rather write a novel or a song based on this insight or write a scientific paper on love than be inspired to embody their insight for personal transformation.

A few of these brahmins, however, eventually get bored with outer expressions of creativity and see the potential of the creative journey for inner exploration. At that point, they hear the call of the evolutionary pressure of purpose from the future and set their creativity toward the study of higher states of consciousness. However, when they attain a higher state of consciousness, they tend to declare themselves enlightened, write a book, and seek fame. So even this ends up as outer creativity of a sort; the product is used an accomplishment of the ego. You see such people abound on the Internet looking for name and fame. The materialist culture continues unabated in the form of spiritual materialism.

In spite of all these barriers, a number of transformation-oriented people had shown up in the agricultural era, and that number has clearly increased in recent times. A modicum of today's people does journey the path of transformation at least partway, a process that the poet John Keats called "soul making." Sri Aurobindo has called this level of mind an illumined mind. As discussed earlier, the people of illumined mind perhaps make up as much as fifteen percent of today's people.

However, as a society, we are nowhere close to developing the next stage of human evolution—the intuitive mind in which the people's mind is preoccupied with giving meaning to the supramental. Why is that?

Emotions have to be properly integrated in our world culture lost in the illusion of rationality while quite incapable of dealing with the "animality" of instinctual emotions. Until the emotional vital mind is fully integrated with the rational, we are stuck, evolutionarily speaking. This integration is the subject of two of my books (Goswami and Pattani, 2022; Goswami and Onisor, 2024).

Will technology help push us up to the next level or do we have to make a true-blue quantum leap of evolution as a species? We will see. This will be the subject of chapter 9.

The Coming of the Intuitive Mind en Masse:
Are we Ready?

After the individuals of the rational era balance negative motions with positive and integrate their emotions with thinking, they are ready to develop the intuitive mind. Right now, 15% of the world's people are interested in such transformation. A small percentage of them may already have done this balancing and integration and are well into the stage of the intuitive mind at the individual level.

How will era of the intuitive mind come about for the entirety of humanity? More and more people will join the 15%, get over their worldview confusion about archetypal values, wake up to higher needs and human potentialities, and decide that without some transformation from their basic ego level of being it is impossible to shift life's emphasis from the outer/physical to the inner/consciousness orientation. These people therefore will not remain intellectually frozen in their adult lives. Instead, they will continue to explore the supramental to discover the archetypal themes of life to a fuller extent. It is likely that most of them will make only mental/ emotional representations of the archetypal themes, but then they will not live it. But surely some will walk their talk and make new brain circuits to make walking this way relatively effortless.

In other words, as the worldview changes, more and more people will fully participate in inner creativity; they will manifest their supramental insights about the contexts of mental and vital functioning in their lives. Since emotions are integrated with reason and meaning processing and feelings and meanings are treated on equal footing with always the proper supramental context as background, these people will live life more or less under the guidance of their intuitions with a preponderance of positive emotions and expansion of consciousness. Will this be the beginning of what Teilhard call bringing heaven on earth? Or taking the Hindu notion of yuga, will this signal the end of kali yuga and the beginning of the ascent of human civilization? Will this be the beginning of the era of the intuitive mind?

There have been such individual beings of intuitive mind in our midst throughout the ages; this has been the case for millennia. They are the people who created and help sustain our past civilizations, some of the rishis of the Upanishads, the great Kabbalists of the Judaic tradition, Lao Tzu, Socrates, Buddha, Jesus, Shankara, Muhammad, Nanak, Ramakrishna, and the like.

Unfortunately, the spiritual/religious or philosophical traditions created by these intuitive minds of the past have been world negating in general. By and large the result has been religions that have succumbed back to rationality and created a tradition of a religious mind that is a combination of rationality and rituals, and not necessarily integrated either. In short, religions have not helped much to raise the status of human mind and civilization beyond rationality.

Fortunately, if we look closer, at least in India, there has been another tradition of spirituality, a tradition that encouraged spirituality staying in the service of the world. It began with the teachings of Sri Krishna in the Bhagavad Gita millennia ago, much misunderstood in the subsequent years but fortunately rediscovered recently by Sri Aurobindo.

More recent philosophers such as the spiritual teacher/author Anirvan, have seen these two paths as distinct: one propounding "world is suffering (*dukkha*)" thus negating the world; this is path of *dukkhabad*—the path of renunciation of the world of suffering. The other path recognizes the potential joys available in the world as we transform and celebrate the joy or happiness. This path is called *anandabad*—it proclaims the positivity of the world.

With quantum science of consciousness to guide us, we are finally able not only to discern between the two traditions but also declare that anandabad—the path of positivity—is the more scientific one for the society although individuals can follow either path. The difference is that the negative path emphasizes the exploration and embodiment only of the archetype of self as the goal of spirituality whereas, the positive path emphasizes all archetypes within the overarching emphasis on the exploration and embodiment of the archetype of wholeness.

What's the difference? In the exploration of the self-archetype, after self-realization, eventually the self dissolves into unity—no boundaries and perfect happiness. This is called liberation and presumably it liberates one from the death-birth-rebirth cycle, but in effect, it is an escape, renunciation of the world.

The exploration of wholeness on the other hand consists of creatively bridging and integrating the dichotomies and create ongoing conflicts to bring suffering that deteriorates the human condition. Whereas the path of self-exploration is renunciation of the world, the path of wholeness is embracing the world while integrating the dichotomies that bring suffering in people's exploration of the world. In self-exploration, the goal is liberation; in the exploration of wholeness, the goal is to become an avatara, dedicated to bring joy to the world.

How the Four Stages of the Human Condition Came About: A Joke I found on the Internet

At the first step toward creating human beings, God created the dog and instructed it thus: "Wait all day by the fence of your house and bark at whoever walks past. For this, I will grant you a life span of thirty years."

The dog said: 'I can't bark that long. Make it fifteen years and I'll give you back fifteen?' God agreed.

At the second step, God created the monkey and told it so: "Jump around, do tricks, make faces, entertain people. For this, how about I give you a twenty-year life span?'

The monkey replied: "Perform monkey tricks for thirty years? No way. How about I give you back fifteen same as the Dog did?' And God said OK.

At the third step, God created the ox and said: 'You must go out into the field and do any which odd job your master finds for you. You have to satisfy him and not yourself. For this I think you deserve the long life of sixty years. Agreed?"

The ox said: "Sixty years of hard work for somebody else? No, thanks. How about I give you back forty and keep twenty?" And the good God agreed.

Finally at the fourth step, God created the human and said: 'Eat, drink, be merry, enjoy your time on the cell phone. For this, I'll give you twenty-five years. Ok?"

The human said: 'Only twenty-five years? I want more. Could you not give me my twenty-five plus all those years my forerunners gave you back?" God agreed, "You asked for it," and gave the human twenty-five human years plus forty ox-years plus 15 monkey years plus another fifteen dog years.

So that is why for our first twenty-five years we eat, drink, play merriment and enjoy our cell phones. For the next forty years we slave in our jobs for other people. For the next fifteen years we do monkey tricks to entertain our grandchildren and to keep doctors away. And for the last fifteen years we sit and bark at everyone and complain, complain, complain.

Are there recent examples of fully or partially transformed people in the pursuit of wholeness, the avataras of wholeness—fully grown or in training—today? people who regularly have peak experiences of creativity and spirituality, who are able to love unconditionally more often than not especially when occasions demand it, who are able to maintain a state of equanimity or environmental independence? Yes. People like Henry Thoreau, Mahatma Gandhi, Rabindranath Tagore, Swami Vivekananda, Annie Besant, Paramahamsa Yogananda, Eleanor Roosevelt, Sri Aurobindo, Abraham Maslow, Desmond Tutu, and the current Dalai Lama seem to fit this role. They by no means can be regarded as perfected beings, yet they were or are able to maintain a sense of humor about their failings.

In my book with physician Valentina Onisor *Quantum Spirituality*, I have shown that living in wholeness is the purpose of human life. In its initiation the adept achieves an intuitive mind as he or she explores and embodies one of the archetypes and begins the journey in wholeness. Part of this journey is to teach others the ways of transformation. This is how they become avatara. At the end of the journey of anandabad, these adepts attain quantum enlightenment, a life in wholeness; even so they never give up teaching.

In this way, human life with many reincarnations is to be looked upon as a learning journey (to learn the supramental archetypes or how and why the world works) that requires many lives. When all archetypal themes of life are discovered and lived, then a person has no use reincarnating again to have another mental life. Therefore, such a person can aspire to be liberated from birth-death-rebirth cycle. That is when dukkhabad becomes relevant.

From Individual Transformation to Societal Transformation

The era of the rational mind has been going on perhaps for 6000 years but not all people think rationally even today, far from it. However, confusion and all, by and large the societies value rational thinking over emotions and intuitions. This is why calling today's societies and the era rational is appropriate.

But how to go from this mindset to a mindset that values intuitions and positive emotions for the entire society? There is reason for optimism that there will be such a shift of mindset. Let me count the signs:

1. The worldview is changing from Newtonian to quantum, slowly yes, but surely.

2. 15% percent of all people, and that is more than a billion people, have been engaging in transformational practices like yoga, chi gong, and meditation.

3. The popularity of the so-called higher education which is in truth an indoctrination into scientific materialism and rational thinking is on the decline.

4. Institutions of transformational education are cropping up, slowing but surely. Should remind you of how liberal education began in the eighteenth century to replace religious dogmatic education.

5. Some of the teachers of transformational education already qualify as avatara; others are in training.

But you know, something even bigger is taking place as well in the purposive movement of consciousness. In the past the changes in the era took place mainly via the advent of new technology: small scale agriculture and big scale agriculture are examples. In the rational era, advent in technology have initiated more and more people toward rational thinking. But recently, advent of technology may be producing a deeper change.

I am talking about the generative artificial Intelligence technology like ChatGPT. Generative AI learns differently from humans, but computer processing is now so fast, that generative AI can cover all existent human knowledge to learn ever more exhaustively, potentially surpassing human capacity.

The question then arises: if AI can do *rational thinking within the known* even better than humans, obviously an AI takeover of all rational and mechanical activities is unavoidable. What would humans do then? Human beings must turn toward thinking either irrationally or *nonrationally* via intuition. The former will bring chaos. However, chances are good that we will go the intuitive creative way.

Contrary to old thinking, quantum science is showing that the potentiality of creative transformation is available to all. To actualize the potentiality, most people need guidance, and such guidance is already available. To avoid chaos, those who can, must join the transformative journey toward developing the intuitive mind in order to usher the new era.

CHAPTER 9

Toward a Nonlocally Connected Society and the Era of the Intuitive Mind

I spoke of the Indian concept of avatara in the last chapter, the great spiritual masters who rescue human societies from chaos and decline. Krishna is recognized to be such an avatara, as is Buddha, as is Socrates, Plato, Lao Tzu, Jesus, Mohammad and many others.

The avataras of the old, inspired people to explore and cultivate the "good" archetypal potentialities; they could inspire people by virtue of their moral authority, they lived and embodied what they taught, *they personified the good, people saw them as gods and goddesses in incarnate bodies, and were inspired to explore the positive.*

Leaders of today even of the 15% because of the sociocultural conditioning of their upbringing become too involved with rational thinking. They do talking about consciousness and alternative worldviews but avoid living the worldview; they do not walk their talk and try to inspire with so-called charisma, not moral authority. They mostly ignore and suppress emotions as they become adult; as adults, they do not bother to (re-)awaken their hearts and develop positive emotions and transform. Naturally, leaders of today will not be able to inspire their followers to transform in the quantum way of wholeness unless…

Unless the leader inspires by his or her own example and teaches the followers how to explore, how to be good, how to make people's good-evil split whole again.

This is what the avataras of the old were able to do because of their authenticity. Quantum science of reincarnation is saying that after many reincarnations that it takes, when a person more or less fully embodies one archetype, he or she becomes an *authentic* avatara of that archetype (they are also called *angsha* or part avatara, or avatara in training). When one explores and embodies the archetype of wholeness (which also entails embodying all the archetypes), she or he is called *purna* (Sanskrit for whole) avatara.

Becoming an avatara comes with a solemn promise of returning over and over to serve humanity—this is the same as the bodhisattva vow, also called the kung yen vow in Chinese Buddhism. The vow is explicitly stated by Sri Krishna in the Bhagavad Gita:

Yada yada hi dharmasya glanir bhabatu Bharata,
Abhyuthanam adharmasya tadatmanam srijamyaham;
Paritranaya sadhunam, vinashaya cha dushcritum,
Dharma sansthpanarthya sambhabami yuge yuge.

Translated in English:

Whenever it becomes impossible for people to follow their dharma,
And wrongfulness enters, I myself appear;
For the protection of the righteous and for the peril of the wrongdoers
For the reestablishment of the right worldview, I return in every era.

Time for a joke. Aliens came to visit earth. No, that Hollywood movie about alien visit is wrong, these aliens did not look like ET; but they did speak English.

Naturally, the pope went to see the aliens. As soon as he got a chance, he asked, "Do you guys know about our savior Jesus Christ?"

The head of the aliens laughed. "Oh yes. We call him JC. He visits us every year."

The pope is puzzled but also excited. Maybe the aliens can give a hint as how to get Jesus back for a second coming. He says, "That is very interesting. He was with us two thousand years ago but never came back. What is you secret?"

"He comes to us for chocolate. Our planet grows too much chocolate and his planet has no chocolate. So, we have a trade agreement with him."

No, we cannot entice avataras by chocolate, but we can with devotion and practices geared toward manifesting and embodying the qualities they had.

Take this idea with you.

> *Avataras—people who embody an archetype's qualities completely become avatara of that archetype in their subsequent incarnations. In the past, they have come and restored order and the pursuit of civilization. There may be avatara-like people today around you and looking for you to transform into one. Are you interested in doing your role, to save human civilization, to restore human evo-devo to its right destiny, to save your planet, our planet?*

Toward a Nonlocally Connected Society, Building Positive Emotional Instincts, and Ushering the Intuitive Era

From the ending of the last chapter, it may seem that we have found the developmental mechanism and processes via which the current chaos and decline of civilization can be overcome. Look around, there are many avatara-likes in recent times by the new scientific definition who have been teaching transformation to so many. It is said, "When the field is ready, people will come." The field is ever more ready, and now not only teachers but also the right integrated worldview is here. And more. Quantum science is giving us new details of the journey of transformation, details not found in the spiritual wisdom literature.

Are people ready to make the required transformation to institute social change *en masse*? Today, the materialist worldview and rational thinking so dominate the academe, the so-called "higher" education so corrupts people who go through it, the lower educated are so dumbed down by information processing, that hardly anyone hears the call of the higher needs in their teens. Most of the 15% wake up at mid-life. That may be enough time to go all the way and become an avatara-apprentice, even angsha-avatara; but it leaves too little time for them to make any big social impact.

We need to expand the scope of our thinking. Is the avatara system enough for our times to restore Dharma? That system worked when people were still largely connected nonlocally. In this mature part of the rational era of high tech, cell phones, and social media, people are anything but nonlocally connected. The "global brain" of local connection is capable of enabling you to talk, but not walk the talk.

The good news is that most of the 15% because of their reincarnational maturity do wake up at midlife. They can engage with the archetypes, embody them to the best of their ability, and transform as much as possible, but let's face it, most of them fall back. Right now, even optimistically, about 1% engage transformation. However, with so much corruption of their minds via the sociocultural milieu they live in, most of the one percent will not be able to develop the moral authority of an avatara, even part-avatara. But equally importantly, under some guidance from the few available avataras and their apprentices, they can build positive emotional brain circuits for themselves, they can also connect, build communities and make collective nonlocal memory of their fruits of transformation. In a few generations, we may have nonlocal connectivity once again! My vision is (along with few luminaries such as Rupert Sheldrake) that should be enough to take us to the next era of the intuitive mind. How? By making positive emotions instinctual.

How Societies Are Built

We know about conglomerates of matter; elementary particles make atoms, atoms make molecules, molecules make bulk macro matter. Our model of societies from materialist reductionist thinking is similar: they see society as conglomerates of intelligent but mechanical machines.

Philosophers called holists add a little twist to this picture. They declare: the whole must be greater than the parts. How? No scientific explanation has ever been given. Holists just assert that some new property emerges that is beyond what the interaction of the components can produce. They posit atoms are something more than a conglomerate of elementary particles; molecules more than a conglomerate of atoms, and so forth. Of course, their real objective is to provide a philosophy for things like life, mind, and consciousness to *emerge* from complex conglomerates of matter.

To be sure, this kind of simple holism is wrong from the start if you insist on being scientific. Reductionists love to point this out: everything we need to know about water, everything that is measurable can be predicted from its ingredient atoms—two hydrogens and one oxygen—and their interaction. If we say that the wetness of water is an emergent property, which seems true on the face, we cannot predict wetness-feel of water from atomic or molecular physics, but on closer look you find that the real explanation of wetness-feel is that it is a result of our conscious interaction with water! *We feel* wetness.

So naturally, holists are making a dubious claim when they make models for the societies of the living as a progression of emergent wholes: from a society of neurons to a colony of bees to a society of humans with brain in the same manner with the same holistic mantra in mind: a society of the living is the result of the interaction of the component living beings; from this interaction arises something more, a society of a bigger whole that governs the earlier but is open at the top.

As the philosopher Ken Wilber has argued in his book *Integral Spirituality*, there is a big difference between everything else and a society of humans—beings with brain with conscious identity. Whereas a colony of bees are often found to act with one center such as when they move as a swarm, human beings in a crowd don't do that; there usually are more centers than one, more often than not.

Quantum science shows that living conglomerates are fundamentally different from nonliving ones. It is in the way the components interact to form the whole. Material conglomerates interact through material interactions, from elementary particles to the universe all the way. Loosely speaking, we say elementary particles make atoms, make molecules, make larger assemblies, but strictly speaking the wholeness of the components are not entirely maintained. In fact, in truth, causally speaking, there are only the elementary particles and their interaction. They just make the universe in such a way that loosely falls into stages of conglomeration.

Also, when we say stars interact to form galaxies, we are not proposing new causal force for stars; it is the same old (in this case gravitational) interactions of elementary particles.

Now compare; do quantum think. A living eukaryote cell is interacting with another such, both are individual organisms. How do they interact to form a colony of cells? Material interaction between molecules of the two? Cells are electrically neutral; but when cells come close together, little fluctuations from neutrality can occur, electrical imbalances, that are a major factor for cellular cohesion. However, such cohesion cannot interfere with the basic integrity of a cell which is defined by a tangled hierarchy and the self of the cell. In other words, a living cell is an irreducible whole.

In quantum science, beside the material local interactions, cells can bond via quantum nonlocality, via consciousness. Consciousness holds them together by programing them to perform orchestrated purposive functions appropriate for the colony; the programming uses morpholiturgical fields and psychophysical parallelism in a way similar to how cells form organs.

However, quantum nonlocality does not help such organisms for interaction with nonliving macroobjects. Thus, for a flock of migrating birds, we find that it is magnets in the bird's brain that guide their directionality via interaction with earth's magnetism. Those magnets in the bird's brain are sensitive to the Earth's magnetic poles.

Why is there one center in such a colony, for example, a queen bee, in a colony of bees? Because the different conditioned programs are organized in a simple hierarchy as in a computer; the center is the analog of the CPU of the computer. The queen bee is the head honcho of the behavior programs that run the individual bee.

What is the difference then between a multiple cell organism or a society of such multiple cell organisms and a mammal or human with distinct organs, especially the brain with self-identity? In the former, there is no central self at the macrolevel. Once programmed, the collapsed experiences are all confined to the single cells. For mammals and human, there is tangled hierarchy and central self at the brain, collapse of possibilities and experiences occur at the macrolevel with which the organism identifies. The tangled hierarchy gives these organisms a new macroscopic integrity—a central self in the brain (and for humans, at a few other chakras as well).

For primitive humans through the vital mind era, nonlocality was responsible for the tribal societies those ancestors of ours lived. The nonlocal connections were through the feelings in the body and consciousness. The body's nonlocally connecting selves of feeling were equal partners of living with the separateness producing thinking self. In the vital mind era, even the thinking selves, via giving meaning to positive feelings promoted nonlocal connectivity. The equation changed, however, with the development of the rational mind, when the thinking-self took over either completely (for the materialists) or enough to create confusion (the religionists) the control of the body.

Thinking rationally is separateness-producing preventing non-locality. Thinking produces complex structures of ego and personality and simple hierarchy. We exchange thoughts via local communication, and with it each other's meaning mostly as information. Nonlocal consciousness does not usually engage. So, interaction via mental information exchange is entirely superficial as when you say, "Howdy," and your friend responds "howdy."

This is why in human societies of the rational era, Wilber's point is well taken; we do move in a crowd with many centers. In fact, we act a lot like "independent separate objects" of Newtonian physics, separate ego-personas.

There is still "we," however, local we. The "we" comes from sociocultural conditioning. We grow up in the same culture and get similar habit patterns. The mirror neurons of the brain have evolved to help us mimic another's behavior; this, too, produces social abilities. The end point is a certain amount of homogeneity that our socio-cultural conditioning gives us; and with that it is much easier to get along with other people of the same culture, identify with that culture. But is this enough for civilization?

Societies we have lived in the past: the Struggle for Civilization

There is a Gandhi story. Gandhi was in England for talks with the British Imperial government about India's attaining freedom. Somebody asked Gandhi, "So what do you think of the Western civilization?" To this, Gandhi is supposed to have said something to the effect, Civilization? I don't see much of it here. More recently, the philosopher Rajani Kanth has written a good expose of the atrocities of Euro-modernism—the metaphysical foundation of the so-called Western civilization—around the world. Read his book, *The End of Euromodernism.*

To be fair though, I think Gandhi's comment applies to other "civilizations" of today's world as well, including the current Indian one which Gandhi himself help build. There isn't much of civilization anywhere!

Truth is, there has not been much of lasting civilization—lasting more than one or two millennia—anywhere on earth during the roughly 4000 years of rational era. The record is clear. Occasionally, some special people of the society—the avataras—take the leadership role; under their leadership, disciples do try to transform and they succeed in making positive emotional brain circuits and develop emotional intelligence. Some even transform all the way. Their disciples follow their leaders and adopt the path of emotional transformation as well. The example of these larger second and even later generations of leaders creates a high amount of respect for this level of being in the society; even the people of tamas, without much spiritual understanding learn to be good and peaceful. The leaders via their moral authority inspire and influence the common people. Civilization thrives. This has happened time and time again when common people have responded to their spiritual leaders and societies *en masse*; in this way, human societies have had their day in the Sun, peaceful and creative epochs of civilization.

Two trends are important to note, however. With time, the civilized epochs have become shorter and shorter; this is the idea of yugas already discussed. In ancient India, Satya yuga was followed by a shorter Treat yuga followed by an even shorter Dwapar yuga.

Then came Kali yuga. Buddha tried to uplift civilization in India from the straitjacket of Brahmanism with great success during his lifetime. As soon as he passed away, however, the nonviolence movement ended, and violence came back in a hurry. Civilization was revived after a while under the guidance another great leader Ashoka who had moral authority but that, too vanished after a century or so.

Following Jesus' teachings, the early Christians tried to establish energies of love as the reigning emotion en masse. But Christianity went nowhere until their leaders co-opted the violent often evil ways of the Roman Empire for spreading their religion. In spirituality, the means and the end cannot be contradictory. Christianity never recovered from this dalliance.

Muhammad founded Islam on the dictum "Allah (God) is merciful", but almost immediately after his death, war broke out between two faction and the division it created is still there.

The second thing to notice is this: no great long-lasting religion has come about since Islam in the 7th century. Not because avataras did not come; Guru Nanak, Sri Chaitanya, St. Francis of Assisi, the more recent Bahaullah, Ramakrishna, and Henry Thoreau, were good examples of really effective teachers, but their effectiveness remained basically local.

The inescapable truth is this: In the entirety of the rational era, we as humanity have never been able to accomplish the nonlocal connectivity to the extent that our ancestors did in the vital era. The force of that connectivity has gradually declined and has basically run out after the founding of Islam.

There is another inescapable truth as well. Our vital era ancestors gave mental/archetypal meaning to the instinctual brain circuits of feelings they inherited from the primates and made the negative mental/archetypal software as part of the humanity's collective unconscious, a universal software that religions call the evil. To be fair, our ancestors knew that there is a "good" counterpart to the evil; they did their best in creating archetypal images of "good" in the forms of gods and angels. Unfortunately, those ancestors did not have the know-how to build much positive emotional brain circuits of the "good".

We have since made great individual strides of transformation including the achievement of emotional intelligence at the individual level in past epochs of civilization; this has never been carried to the masses though. With little built-in positivity, the negativity of the brain is too much for ordinary people to be steady-forth in positive behavior.

So, it has been much like the Sisyphus myth played out repeatedly for the entire humanity. We raise the boulder of civilization uphill in some locality only for it to fall down again after a while; and then we try again somewhere else at some other time but always with the same end result.

The immature conclusion is of course the one currently adopted by the USA. The decline of civilization happens because of the invasion of outsiders who are uncivilized. To save civilization, you never give up your military might. That will protect your civilization from external invasion, the enemy without. But the real enemy is within; we are our real enemy.

USA in the nineteen sixties and seventies had the beginnings of a renaissance towards civilization. There was a "greening" of America. But the color rapidly changed to brown with the advent of scientific materialism and the religious backlash against it. Today, USA unquestionably remains the strongest in the whole world militarily, but where is civilization?

Somebody said, "Man has learned to fly like a bird in the sky; man has also learned to swim like a fish in the ocean; but alas! Man has yet to learn to walk like a human being on this earth."

I will put the anguish in another way with apologies to the great poet Alexander Pope:

> All that archetypal good
> Lies hidden in the night;
> Where is the metaphorical magic
> That will bring it to light?

Clannishness and Tribal Mind-sets

Even in rational societies, nonlocality does not go away entirely. People can go to churches; neighbors can interact in neighborhood activities; parents can interact with other parents in children's school activities. Young people go to schools and colleges. In sum, people in a community can get correlated; nonlocal connection can then happen within the community because feelings come into the picture; with it comes the tribal mind-set—clannishness. However, for people outside the tribe—strangers—due to the absence of local or nonlocal connection, feelings are dominated by fear and lead to contracted consciousness, not nonlocal expansiveness and inclusivity. This is why almost every human culture has a pejorative name for foreigners.

Often this prejudice against "foreigners" even carries over to religions growing in an isolated culture. So, Hindus have a pejorative word for Muslims, *mlechhya*. Muslims likewise regard Hindus as *"Kafirs."*

When I was growing up in undivided India with my Muslim neighbor as part of the same culture and came across these concepts, I was puzzled. "My Muslim friend looks the same, talks the same." But an older Hindu friend was quick to point out, "Oh, but he eats garlic! Yuk." Hindu brahmins in Bengal did not eat garlic at the time.

Rationality in humans cannot dominate the instinctual negative emotions; in fact, it is the other way around. People of a community easily pick up each other's base emotions such as fear through their mirror neurons. This gives rise to en-masse racism, sexism, religious bigotry, etc. that charismatic leaders can use to rile up their followers against "foreigners." Sounds familiar?

As a young boy in British India, I grew up with the Bengali word *saheb* for white people, the literal meaning of the word is "lord and master." In British India, that was only natural; white people governed us, they *must be* superior. Although I did not know any white man or woman personally, from my readings and from all the signs around me, the concept made sense. Yet it conflicted with the deep sense of nationalism that our political leaders and freedom fighters were influencing me with, *I am not inferior.*

India became independent when I was ten, but I had to wait till I was a PhD student before I actually met my first white person who quickly became a friend, a Polish visiting physicist. He was fascinated that I never drank alcohol in my entire life and started talking about the virtues of drinking. In Poland he said a man's caliber is entirely judged on his drinking ability; "the leader is the one who can outdrink all the others under the table," he said. He then explained, "See, you sit around a table and start drinking. One by one, as people reach their limit, they have to go under the table. The last one on the table is the best man."

Few years later, I was in USA then for three years, was no longer a stranger to alcohol, and was attending a conference. In the evening, I went out with my roommates, two fellow physicists, one Chinese, one British. That was the first time I got drunk; somehow, I knew that I had to slide under the table, and my friends were pulling me up. Eventually, they must have taken me back to the conference hotel. I woke up in the middle of the night quite sober; my friends returned after a while, and now *they* were drunk; lo and behold, the Britisher was "swimming" on the carpet. And I thought, "He is no different than me." It was a revelation; it was also a liberation.

This is the thing. One has to be intimate with a "different another" in order to penetrate the deep layers of bigotry circuits in the brain. The pretend behavior that materialists preach is superficial; in their heart people remain hateful, mistrustful of "foreigners," people who are *different*. This is why today in America and elsewhere there is so much mistrust and hate among people.

Take this idea then with you:

> *Whereas vital era societies used locality as well as nonlocality as they bonded, mental/rational societies use locality only because people solely live at the superficial ego level. Mental/rational societies tend to be individualistic, but that too is a mirage. There can be clannishness at the suppressed emotional level acting subliminally for people of same color and creed, sex and religion, and worldview, an unconscious but equally powerful clannishness that excludes people of different color or creed or sex or religion or worldview. This*

is responsible for such persistent societal proclivities as racism, sexism, casteism, religious prejudice, homophobia, and worldview polarization.

Transformation

At the next stage of our evolution which we should be ready to welcome right now, the mind should become capable of giving meaning to intuition as its major preoccupation; the human personality can then become more and more intuition-based. When we are intuiting, we are momentarily in the quantum self. If we do this often enough, that is, operate on intuition more often than not, then once again, nonlocality will dominate our interpersonal interactions opening door to more positive evolution. With nonlocality operating, unity consciousness operating between us, we can work toward a tangled hierarchical relationship with other humans. As you know tangled hierarchy leads to a self-identity, a true nonlocally acting center, bonding together all the individual centers. *This is a "we" at a higher level of consciousness.*

The challenge of the twenty-first century human is this: how can each of us help to bring about this tangled hierarchical *we* of a quantum society but without losing our individuality either? Please note that this is not anything like the artificial intelligence called "Borg" of science fiction episodes from *Star Trek: the next Generation.* Having tangled hierarchical relationships some of the time does not rid of our individuality at other times, in other relationships, at all.

But of course, there are currently huge barriers against the evolution of the intuitive mind en masse. The foremost one is brain's built-in bias toward negativity. Research shows that it takes five happy events (events that expand our consciousness) to negate the effect of one unhappy one. An individual maybe able to transform and develop positive emotional brain circuits to balance the negative. How does she transfer her change to the masses who will not, cannot (they are no-Can do people, remember?), engage in transformational practices?

One way is something similar to the phenomenon of directed mutation mentioned earlier. *When threatened with severe existential crisis, humanity may become nonlocally connected again in a hurry.* Then it is just a matter of evo-devo.

I discussed quantum evo-devo in chapter 7—Darwinian genetic evolution (propagating through heredity) and Lamarckian epigenetic development of form and function (propagated by nonlocal memory and group reincarnation) hand in hand, quantum mechanisms of crucial importance in both. *This is how Lamarckian concepts of epigenetic developmental evolution are integrated with Darwinian concepts of genetic evolution in the quantum theory of creative quantum evo-devo.*

Below I show that quantum evo-devo is the explanation of the phenomenon of instincts. Instincts are also additions to the universal vital software of a species. Why is this important? Because the developmental change we are looking for that will take us to the next phase of mental evo-devo is precisely the development of new *positive emotional* instincts.

The Nature of Instincts

The negative emotional brain circuits that human brains are burdened with are widely recognized as instinctual in origin—the feelings associated are called instincts common to animals as well. What are some other examples of instincts? How do we know that they are epigenetically inherited traits?

A humming bird looks at a red scarf and becomes terribly agitated. This is an instinctual feeling that it is responding to.

You may side with traditional biologists to think that such instincts are programed in animals including us via the genetic evolution, but you would be wrong. When you look at some of the complex instinctual behavior of animals, you will rid of all scintillas of doubt and conclude that gene variation and natural selection are not a viable explanation of instincts. *Instinct have too much of the memory characteristics of software; they are the result of software development.*

Consider a few more examples of instincts to make the point clear. A chick still inside the eggshell gets out by pecking. Genetic codes are basically instructions to build proteins for organs to function in a general way, for example for the chick to peck at the shell at random from inside. But the chick seems to know not to peck at random but just at the part where there is an air space. You cannot call this learned behavior of the usual familiar local kind—nurture—learned during growing up either because the behavior shows up even before chicks see daylight. The chicks learn where to peck via collective nonlocal memory (of nonlocal group consciousness) while still inside the eggshell.

If you are still skeptic, consider the following tale of instinct told by the geneticist Henry Fabre. Beetles lay eggs within the body of a tree. The eggs hatch and the sluggish larvae called grub spend *three whole years* within the tree burrowing through the wood. When they emerge from the tree at last, they build a nest. The complicated nature of the nest should leave no doubt that the instructions could not have come from any genetic change.

Conclusion: Instincts are Lamarckian in evolutionary origin; they are examples of quantum evo-devo in action to be specific.

A Viable Quantum Science of Animal Instincts

Consciousness can use the morpholiturgical fields potentialities of the vital body in three ways: 1) fundamental creativity, 2) situational creativity and 3) inertia, no creativity—just some random shuffling of the existing repertoire.

If the adjustment required is so small that one can get by without making changes, then conditioning prevails, the random shuffling needed is automatic. If the adjustment to the environment requires a moderate change, a solution may be found creatively using the existent repertoire of the morpholiturgical field potentialities to find a new combination—situational creativity—also automatic. If on the other hand a major change is called for, fundamental creativity must come into play. This is not automatic.

For nonhuman animals, at first glance the learned new habits of situational creativity may seem to be individual, but the nonlocality of the memory (of learning) involved and nonlocal connectivity in the vital arena assures that once one member of a community learns a habit, the nonlocal nature of the memory makes it easily accessible to other members of the community as well; it facilitates the learning for the whole community, a point first noted by the biologist Rupert Sheldrake. In this way, the learning can spread universally for a small group of *correlated* animals quite rapidly, and if these groups can intermix because they live in a colony (which provides local connection and correlation), eventually for the entire colony. If the colonies are isolated and cannot get correlated, the spreading can still take place via reincarnation of the surviving group quantum monads of individual colonies taking longer time; reincarnation will randomly shuffle and mix up the individual members of different isolated colonies. Eventually, the learning will spread throughout the whole species in a few generations.

This universal species nonlocal memory constitutes the collective unconscious of each individual species including human beings of the vital mind era.

Instincts are then hereditary developmental information transfer through creatively learned abilities for using the morpholiturgical fields during early development; these abilities are triggered in the embryo and the early childhood when needed; this leads to the development of instinctual software of the individuals to be added to other universal software. The nonlocal memory of learned propensities is propagated by Group consciousness and Group reincarnation. Keep this insight with you.

For humans in the rational era, the surviving quantum monad becomes individual; the dynamics of collective memory making becomes more complicated.

The Evo-devo of the Triune Brain

Because of fossil gaps and lack of clear intermediates, there is a lot of controversy about the evolution of mammals and reptiles. On the basis of the clear triune layered structure found in the human brain, neuroscientist Paul MacLean (1990) suggested that first came the reptiles with reptilian or hind brain, then evolved mammals with an additional mid or limbic brain, and finally, the cortex evolved in higher primates and especially humans (fig. 11). Biologists debate this theory because mammal-like reptiles have been found in the fossil data which they suggest, point to a common ancestor of both mammals and reptiles. Also, researchers have found evidence of aspects of mammalian brain in some reptiles.

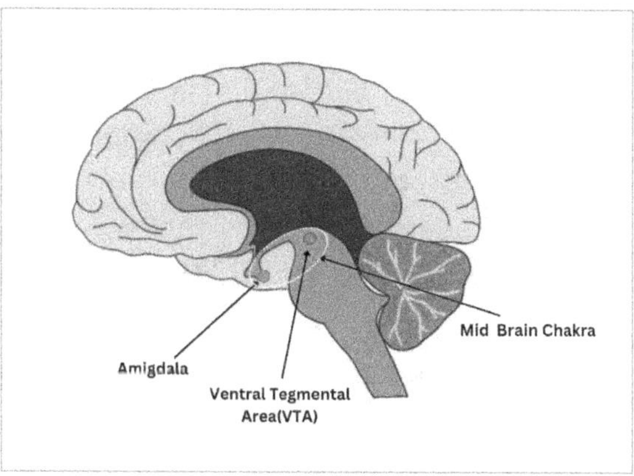

Fig. 11. The triune brain. The figure also shows the midbrain chakra and the organs there.

Neuroscientists of today like the Mclean model, however. The quantum evo-devo creative evolution model solves the dilemma by the observation that what are called early mammals, mammal-like reptiles, or reptiles having mammalian features are all "failed" intermediates within the fossil gap that must exist between reptiles and mammals.

It is also generally agreed that further evolution of the mammals really took off in many new directions at the onset of the Cenozoic geological era which began with the catastrophic extinction of the dinosaurs. It follows that the quantum evolution from reptiles to mammals took place in the millions of years before this event when dinosaurs ruled the earth. This in order for smaller reptiles to survive the onslaught of the dinosaurs.

Imagine the evolutionary theater when there was an environmental catastrophe in the form of huge dinosaurs ruling the earth and rapidly exhausting its food supply. The most advanced animal species of the time is the reptile. Reptilians have a primitive reptilian brain consisting of the innermost part of the current human brain at the back of the brain—brainstem and cerebellum. They also have a nervous system that connects their primitive reptilian brain to the rest of the body's organs. This is not adequate for small reptiles to meet the survival threat from the larger reptiles—the huge dinosaurs.

Still the small reptiles try hard. There is both intense intention and urgency of survival necessity that are consonant with the purposive evolutionary movement of consciousness. Suppose under these conditions, all the necessary new genetic potentiality for a new species with a new mid-brain addition to the old reptilian brain becomes available through gene mutations for consciousness to choose. A quantum leap takes place and some of these smaller reptiles evolve into mammals. The new species has a new layer of the brain that we call limbic or midbrain that would be capable of self-identity (see chapter 7) that gives them adequate new creative intelligence to survive.

This evolution of the self in the mammalian brain would serve survival need in the following way. It would enable consciousness to collapse at the brain nonlocally communicated vital energy of the dreadful stimulus of the dinosaur at the chakras in the body via the neuronal connection at the physical level; in effect, this would amount to a brain take-over of the control of the organs of the body providing the animal much quicker coordinated response. This would also necessitate some additions to the nervous system for carrying out the responses. In this way, the new limbic brain was equipped with the capacity of rapid flight-fight responses to stimuli. This assured the survival of the mammals against the dinosaurs.

Out of these experiences, the new mammals now made shared personal software and universal instincts via the process outlined above.

In this way, the brain-takeover of the first four chakras develops first, personal but nonlocally shared software that enabled instinctual responses; for example, the flight-fight-feeding-f-king responses, the four F's as they are jokingly called by neuroscientists; *the last two F's serve survival necessity no doubt, but they also bring pleasure and pleasure is a doorway to higher need.*

Hunger and sexual drive (the last two f's—feeding and f-king of instinctual behavior) are also built in as brain circuits in the new mammalian brain. So, additionally, these circuits opened the animal to a new vista of fulfilment of higher needs!

A note in passing. A meteor shower was a major factor in wiping out the dinosaurs so that mammals could start evolving explosively. From a consciousness point of view the meteor shower was not a mere coincidence but an act of synchronicity—meaningful coincidence. When we actively try to change and do our best, help comes in the form of synchronicities, discovered by Carl Jung: this is a law of consciousness science.

How Mammalian instincts of Feeling Become Emotional Instincts of Feeling plus Thinking in Humans

The above is a rough sketch of how the human midbrain gets the feeling component of its instinctual negative emotional circuits that we humans experience. Animals you will notice have instinctual feelings, not emotions. Feelings arise, feelings go away; they do not linger on as emotional thoughts do nor can they be recalled easily. In this way, a deer cub will be afraid of a lion cub when the latter is hungry; but the same deer cub is found to play with the lion cub when the latter is not hungry. How can the deer cub tell if the lion cub is hungry or not? Via nonlocal communication with feelings!

All mammals have the capacity of vision and a midbrain organ called superior colliculus to integrate what it is seeing and monitoring eye movements. But this is purely physical. Seeing a lion cub does not necessarily make a dear cub afraid as noted above but grown-up dears to not play with lions ever. Why?

The mammalian brain developed primitive vision capacity of the superior colliculus and enabled the organism to locally receive dangerous stimuli. The adult deer develops feeling software made with the help of the memory making capacity of the small limbic cortex. The adult dear responds to the local stimuli with its instinctual feeling software; it loses the nonlocal capacity.

The mental component of instinctual emotions requires the evolution of the cortex with the capacity of making maps of the meaning-processing mind. Mind gives meaning to the feelings.

During the so-called era of garden agriculture, mind gave meaning to the instinctual feelings (see chapter 8). Men and women worked together in their garden; naturally their feelings would flare up and it was necessary for consciousness and mind to engage with those feelings and sort out their meaning. Feeling plus meaning equals to what we call emotions.

Since feelings associated with survival, generally speaking, contract consciousness, mind interpreted the instinctual feelings as mostly negative. To repeat, our ancestors in the era of the garden agriculture were tribal and nonlocally connected; the memories of learning they made when giving meaning to feelings aroused by stimuli were shared nonlocal memory. Naturally, these mostly negative emotional memory became part of a universal repertoire, the human collective unconscious.

In this way, our ancestors developed the collective unconscious of these capacities of responding to stressful stimuli with negative emotions. This was the genesis of the negative emotional brain circuits of fear, lust, anger, violence, competitiveness, domination, etc.

Fulfilling the drives of the two instinctual F's—feeding and sex—was interpreted by the mind as pleasurable, especially because of the association of the molecules of euphoria—the dopamine molecules for example. Some of the negative emotions mentioned above arose in connection with pleasure-seeking.

Where do the positive emotional instincts of the collective unconscious come from? Our brilliant ancestors did something more than engage in survival. They had gotten into intuitions and archetypes to satisfy higher needs and had started giving meaning to these experiences as well. Since they still cognized with feeling quite a bit with likes and dislikes, they applied that two-fold classification to archetypes and split each archetype into a dichotomy. For example, the archetype of goodness was split into the duality of good and evil, separate.

Finally, they associated the negative meanings given to the instinctual animal feelings as the archetype of evil; they even personified that archetype: under one label—the devil—in the West and many forms of *asura* or *danava* in India.

They personified the good archetypes too: in the form of angels in the West and *devas* in India. However, our ancestors did not explore and embody the devas much; they got stuck in excessive pleasure-seeking. This means that most of the archetype of good remained as potentialities.

In this way, unfortunately, as neuroscientists say, the brain is programmed to be five times more negative than positive. This is at least in part because humans as a species did not quite come to terms with their higher needs during the vital mind era. The saving grace was that relationships were more or less tangled hierarchical at both personal and societal level; there is now clear evidence that the societies were egalitarian suggesting tangled hierarchy.

Today, the evil of our collective unconscious influences us much more than the good. *You can see what that does to people if they do not explore the quantum potentialities of good further and transform.*

Finally, and very importantly, in the rational mind era, the cortical thinking self not only took over the control of the functions of the organs of the body but also that of the midbrain. That is how, we the victims of the rational mind, especially the males, are virtually unaware of these other three selves we have at the heart, at the navel, and at the midbrain. Today it is customary for neuroscientists to call the negative emotional brain circuits "unconscious memory."

Keep these important teachings of quantum science in the forefront of your awareness when you work on transformation:

> *Quantum science has given us a clear understanding of the instinctual emotions and how they came about and how they are weighed so much toward negativity.*

> *The materialist theory of emotions—that emotions are midbrain's response to suitable stimuli—may be superficially true for humans, but this theory cannot explain animal behavior (why a deer cub does not always get fearful of a lion cub; if the lion is not hungry, they even play together).*

> *Quantum science makes it clear how we have forgotten the ownership of the rest of our body and the selves therein. Now parts of who each one of us really is, truly are strangers in our body, these parts of who we are being held captive of our thinking self.*

> *If we want to change our base-level of human operation, we have got to learn how to regain conscious ownership of the rest of our selves in the body; and*

> *We have got to learn how to be nonlocally connected again so that we can make new instincts of positive emotions.*

Directed Evo-devo: Directing Our Future

Even with some transformation—having positive emotional brain circuits and the character to use this positive software—of a few human beings who would provide leadership, is there any lasting gain for humanity as a whole? Sure, this is just a beginning, but the prospect of such mass gain does not look bright. And yet remember: those who transformed all the way after discovering Dharma in a new way in previous times, are the spiritual masters—Rama, Krishna, Lao Tzu, Socrates, Buddha, Jesus, Muhammad—who originated our great spiritual traditions that each became the harbinger of great uptake in civilization for a while; why should this time be any different?

Agreed, the spiritual traditions degenerated into religions of dogma, and the role of religions in civilizing the world has always been dubious at best, and now in this science-dominated era any such role for religion is out of the question. Agreed that Newton's and Darwin's great discoveries did end up in a religion-like revival of materialism in our society and that is partly responsible for the decline of civilization around us no doubt. So powerful was this movement toward materialism that it diverted even the initial big bang impact of quantum physics into a mere whimper for decades.

But now that quantum science within the primacy of consciousness is finally integrating science and spirituality and giving us a new scientific way for transformation, is there renewed hope for a revival of civilization once again?

The finale of human mental development, Teilhard the Chardin called it the omega point, is for humans to develop the capacity for representing the supramental in their physical body—higher vital and mental software. How and when is this going to come about? It is a little premature to engage with that question. A more pertinent question is to ask this, How will nonlocal connectivity of humanity come about once again? Do we wait for an existential crisis and hope that nonlocality will descend even if we don't do anything (something like the idea of second coming in Christianity and the rising of Kalki avatara in Hinduism)? Or can we do something now toward hastening our evo-devo, *given that we have discovered the key to quantum spirituality?* In other words, can we direct our evo-devo as in the phenomenon of directed mutation?

Is there a Hundredth Monkey Phenomenon?

Decades ago, in the nineteen seventies, there was a bit of purported anthropological research that caused quite a stir in the people who identified with the coming New Age. That included me.

Apparently, researchers studied some monkeys in a Japanese island found that monkeys there have learned a new trick; they have all learned to wash sweet potatoes in sea water before eating. Apparently, it started with a monkey mother accidentally doing it; preferring the new salty taste not only did she make this into a ritual for herself but also taught her children the same ritual. The children taught other children, and very quickly, too quickly, the ritual was being practiced not only in that island of origination but also in nearby islands as well, separated by vast expanses of water with no local connection. This later had to be an act of nonlocal communication, this was the conclusion.

The popular book that made this experiment famous called the phenomenon by the evocative name The Hundredth Monkey Syndrome, recognizing a threshold effect in all this both local and nonlocal propagation of a learned habit.

I was quite enthusiastically publicizing this as an example of nonlocality as I went through my own discovery of consciousness in quantum physics, assimilating Sheldrake's work, etc. This was the late eighties. One day, I had my anthropology colleague in the audience and after my talk, she challenged me:

"Amit, that experiment never happened the way you described it."

I was surprised at the accusation. "What do you mean, Gerry?"

"Somebody misled you. I will bring you the original paper."

And indeed, she did and to my chagrin, I found that she was right when I read it. The original scientific data never claimed to have demonstrated that the habit of washing sweet potatoes in salt water spread in other islands. What was learned stayed confined in the island of origin. End of story.

Of course, I soon figured out what was wrong with all of that New Age wishful thinking. Sheldrake and others were all assuming nonlocality, nonlocal morpholiturgical fields doing the spreading, but of course, that kind of thinking is dualistic. To apply *quantum nonlocality requires local correlation first*; in the least mutual intentions. Both were lacking for monkeys in isolated islands.

On the other hand, what if we do correlate people—the Internet, our "global brain" has made this so simple—before engaging group creativity in developing emotional intelligence? Then indeed the phenomenon of the hundred monkeys can be resurrected!

Of course, it is still a far-out proposal, to expect that humanity, in this intellectual era, can hold motivation in large groups and that long! It may take generations. Forget it? Not necessarily. Let me remind you of directed mutation once again.

We don't need directed mutation; we need something similar, directed quantum evo-devo. We need the evolutionary fervor to spread across humanity in a locally correlated way, so that humanity, at least in small groups can join hands, virtually and metaphorically at least, to engage into creatively developing emotional intelligence all together. We also propagate quantum activism, we start transformative education everywhere, we do whatever it takes.

The bacteria were motivated by survival needs; they were facing extinction. We are too facing a survival challenge, but maybe not with extinction. So, the bacteria succeeded to survive in a matter of a mere two days; we will likely take much longer to achieve our goal.

Although the work ahead of us will be transformative work toward a higher state of consciousness, the ongoing crisis for survival will keep us vigilant; they will act as added motivation.

When it comes to survival, we feel the needs in our body, and this cannot be denied as evolutionary data shows. We need this to go beyond this kali yuga of rational madness. We will have to wake up to group or community consciousness and engage group creativity. Then only like caterpillars transforming into butterflies, we shall fly with the ability of emotion management in the skies of the supramental.

Evolving Instinctual Brain Circuits of Positive Emotions

What does it mean to be fully rational? Human beings began with a tremendous leap in the capacity for making representations of the mind. So clearly, a major privilege of being human is in developing the rational capacity fully. But this is not easy; the so-called rational people of today suppress their negative emotions which affect their actions subliminally; this is why see so much irrational behavior on their part, such as sexual harassment, such as propagating the idea that we are robots.

The philosopher Ken Wilber emphasizes that at each level of the great chain of being, we need to achieve a full closure of the achievements of the previous level. We also need to have an openness toward the next level.

To me, closure of the vital mind era means this: we have to accomplish a fully emotionally balanced brain to respond to stimuli; for that we need the collective unconscious equally balanced between negative and positive—evil and good. Only then can we be said to have fulfilled our full potential for rationality.

It is the nonlocal connectivity of our ancestors that created the meager amount of positive emotional instincts in us—altruism, maternal instinct, and even some automatic capacity for romantic love in our teens; starting from the mere beginning of self-domestication, it took them many millennia. We likewise have to develop nonlocal connectivity starting from a beginning of a quantum science of nonlocal connectivity and transformation.

The phenomenon of hundred monkeys did not spread over isolated islands because of the lack of local correlation; this problem is a no-brainer today because of the worldwide web. So, we need to start small Internet communities and transform together to create shared positive emotional propensity.

These communities will grow and eventually the transformation—the positive emotional propensities—will be shared by everyone via reincarnation in the way previously explained. This will finish the job our ancestors valiantly begun, and there will be closure. This will also help restore the nonlocal connectivity that we lose when rationality sells itself to the evil—the Faustian bargain. When evil is balanced by the good, people's rationality becomes open again toward intuition.

Let's Intend: We all who can, must participate in these four steps.

So, using the power of this intention, how do we evolve the instinctual patterns of our species to include positive emotions?

1. *We consciously develop group consciousness in our transformative journey and build positive emotional software for the entire group.*

2. *We participate in transformational work on emotions in bigger and bigger groups. This we do first, in two-way relationship, intimate relationships—lovers, therapist-client, parent-child etc. Then we work in families. Next, we work in groups such as in a business setting or in an educational context. Always, we use the power of intention to propagate the effect even outside our expanding groups. In a matter of a few generations, our brain circuits of positive emotions are inherited by the bulk of the population via reincarnational transfer of positive emotional universal software across the entire species.*

3. *All humanity then sooner or later, learns to associate the archetypal good to give meaning to their positive experience.*

This is the final step in developing emotional intelligence for the entire species.

In this way, babies of a future generation will be born with propensities of making love and goodness circuits in their brain like never before; with their natural access to the quantum self, they will easily make love and goodness circuits to balance their violence and domination circuits. It is up to us to build quantum societies that will educate our children to use these emotional circuits with balance and harmony.

Our research following Carl Jung's original breakthrough indicates that our wise ancestors of the vital mind era already initiated and codified the process discussed above. As mentioned before, they called it hero's journey consisting of the creative exploration and embodiment of the archetypes power along with the archetype of love and goodness. Love civilizes the use of power and teaches you how to share your power with one other. The exploration of goodness allows you to share your power with others and thus empower the entire society.

We have made some progress. Now our society is ready for more ways to civilize the lust for power and the greed for abundance via the exploration of truth, justice, and wholeness. Truth will replace the materialist pretensiveness with authenticity; justice will bring equal opportunity to everyone, the essential element of true participatory democracy and capitalism; wholeness will help resolve all conflicts with negotiation and persuasion and lead to lasting world peace.

We have done this before; we will do this again.

REFERENCES AND FURTHER READING

Aspect, A., J. Dalibard, and G. Roger. 1982. Experimental test of Bell's inequalities with time varying analyzers. *Physical Review Letters* 49:1804–6.

Aurobindo. 1996. *The Life Divine.* Pondicherry, India: Sri Aurobindo Ashram.

Beauregard, M. (2008). *The Spiritual Brain.* N.Y.: Harper-Collins.

Birch, C. (1999). *Biology and the Riddle of Life.* Sidney, Australia: Univ. of New South Wales Press.

Bohm, D. (1980). *Wholeness and Implicate Order.* London: Rutledge & Kegan Paul.

Brough, J. (1958). Time and evolution. In *Studies in Fossil Vertebrates,* ed. T. S. Westoll. London: University of London/Athlone Press.

Brown, G. S. (1977). *Laws of Form.* New York: Dutton.

Cairns, J., J. Overbaugh, and J. H. Miller. (1988). The origin of mutants. *Nature Vol.* 335, pp. 142–45.

Carroll, S. B. (2005). *Endless Forms Most Beautiful.* New York: Norton.

Chalmers, D. (1998). *The Conscious Mind.* Oxford: Oxford University Press.

Chopra, D. (1989). *Quantum Healing.* N.Y.: Bantam.

Cremo, M. A. (2003). *Human devolution: a Vedic Account.* L. A.: Torchlight Publishing.

Darwin, C. (1859). *On the Origin of Species by Means of Natural Selection or the Preservation of Favored Races in the Struggle for Life.* London: Murray.

Dawkins, R. 1976. *The Selfish Gene.* New York: Oxford Univ. Press.

Dembski, W. A. (2002). *No Free Lunch: Why Specified Complexity Cannot Be Purchased without Intelligence.* New York: Rowman & Littlefield.

Dossey, L. (1982). *Space, Time, and Medicine*. Shambhala.

Eldredge, N. (1985). *Time Frames*. New York: Simon & Schuster.

Eldredge, N., and Gould, S. J. (1972). Punctuated equilibria: An alternative to phyletic gradualism. In *Models of Paleontology*, ed. T. J. M. Schopf, 82–115. SanFrancisco: Freeman.

Elsasser, W. M. 1981. Principles of a new biological theory: A summary. *Journal of Theoretical Biology* vol. 89, pp.131–50. (1982). The other side of molecular biology. *Journal of Theoretical Biology* vol. 96, pp. 67–76.

Emoto, M. (2004). *The Hidden Messages of Water*. N.Y.: Atria Books.

Goldschmidt, R. (1952). Evolution as viewed by one geneticist. *American Scientist* 40:84–123.

Goleman, D. 1995. *Emotional Intelligence*. New York: Bantam.

Goswami, A. (1989). The idealist interpretation of quantum mechanics. *Physics Essays* 2:385–400. (1993). *The Self-Aware Universe*: How Consciousness Creates the Material World. New York: Tarcher/Putnam. (1994). Science within Consciousness: Developing a Science Based on the Primacy of Consciousness, IONS Research Report CP-7. Sausalito, CA: Institute of Noetic Sciences. (1997a). Consciousness and biological order: Toward a quantum theory of life and evolution. *Integrative Physiological and Behavioral Science* vol. 32, pp. 75–89. 1997b. A quantum explanation of Sheldrake's morphic resonance. In *Rupert Sheldrake in der Diskussion. Das Wagnis einer neuen Wissenschaft des Lebens* (Gebundene Ausgabe) [*Scientists discuss Sheldrake's theory about morphogenetic fields*], ed. H. P. Durr and F. T. Gottwald. Munich: Scherz. (2001). *Physics of the Soul*. Charlottesville, VA: Hampton Roads. 2004). *The Quantum Doctor*. Charlottesville, VA: Hampton Roads. 2008). *Creative Evolution*. Wheaton, IL: Quest Books. (2014). *Quantum Creativity*. N.Y.: Hay House. (2018). *The Everything Answer Book*. Charlottsville, VA: Hampton Roads. (2024). Quantum measurement theory: the science behind consciousness and experiences. *Journal of Quantum Science of Consciousness*, vol. 2, issue 1, pp. 10-24.

Goswami, A., and Onisor, R. V. (2020). *Quantum Spirituality*. Eugene, OR: Luminaire Press.

(2022). *The Quantum Brain*. Eugene, OR: Luminaire Press.

(2023). *Quantum Integrative Medicine*. N.Y.: Monkfish.

(2024). *The Return of the Archetypes*. Eugene, OR: Luminaire Press.

(2024). *The Awakening of Higher Intelligence*. Eugene, OR: Luminaire Press.

Goswami, A. and Pattani, S. (2021). *Quantum Psychology and Science of Happiness*. Eugene, OR: Luminaire Press.

Goswami, A., and D. Todd. (1997). Is there conscious choice in directed mutation, phenocopies and related phenomena? An answer based on quantum measurement theory. *Integrative Physiological and Behavioral Science* Vol. 32, pp. 132–42.

Gould, S. (1980). Is a new and general theory of evolution emerging? *Paleobiology Vol. 6, pp. 19–30.*

Gould, S. J., and N. Eldredge. (1977). Punctuated equilibria; the tempo and mode of evolution reconsidered. *Paleobiology* vol. 3, pp. 115–151.

Grad, B. (1964). A telekinetic effect on plant growth. *International Journal of Parapsychology* vol. 6, pp. 472–98.

(1965). Some biological effects of "laying-on of hands": A review of experiments with animals and plants. *Journal of the American Society for Psychical Research* vol. 59, pp. 95–127.

Graeber and Wengrow, (2021). *The Dawn of Everything*. N.Y.: Penguin.

Grinberg-Zylberbaum, J., M. Delaflor, L. Attie, and A. Goswami. (1994). Einstein Podolsky Rosen paradox in the human brain: The transferred potential. *Physics Essays* vol. 7, pp. 422–8.

Grof, S. (1998). *The Cosmic Game*. Albany, N.Y.: SUNY Press.

Gruber, H. (1981). *Darwin on Man: A Psychological Study of Scientific Creativity*. Chicago: University of Chicago Press.

Hitching, F. (1982). *The Neck of the Giraffe*. New York: New American Library.

Harari, Y. (2014). *Sapiens: A brief History of Humankind*. London: Harvill Seeker.

Hawkins, D. (2020). *The Maps of Consciousness Explained*. N.Y.: Hay House.

Jung, C. G. (1971). *The Portable Jung*, ed. J. Campbell. New York: Viking.

Kanth, Rajani. *Farewell to Modernism*.

Kimura, M. (1983). *The Neutral Theory of Molecular Evolution*. Cambridge, MA: Cambridge Univ. Press.

Kirschner, M. W., and J. C. Gerhart. (2005). *The Plausibility of Life*. New Haven, CT: Yale Univ. Press.

Laszlo, E. (2004). *Science and the Akashic Field*. Rochester, VT: Inner Traditions.

Lipton, B. (2005). *The Biology of Belief*. Santa Rosa, CA: Mountain of Love / Elite Books.

MacGregor, G. (1978). *Reincarnation in Christianity*. Wheaton, IL: Theosophical Publishing House, Quest Books.

Maclean, P. (1990). *The Triune brain in Evolution*. N.Y.: Plenum.

May, R. (1994). *The Courage to Create*. N.Y.: Norton.Moser, J. S., Schroder, H. S., Heeter, C., Moran, T. P., Yu-Hao, L. (2011). Mind your errors: evidence for a neural mechanism linking growth mind-set to adaptive posterior adjustments. *Psychol Sci.* vol. 22.

Myers, C. A., Wang, C., Black, J. M., Bugescu, N., Hoeft, F. (2016). The matter of motivation: Striatal resting-state connectivity is dissociable between grit and growth mindset. *Soc Cogn Affect Neurosci*, Vol 11, pp. 1521-1527.

Maslow, A. H. (1971). *The Farther Reaches of Human Nature*. New York: Viking.

Mitchell, M., and A. Goswami. (1992). Quantum mechanics for observer systems. *Physics Essays* vol 5, pp. 525–9.

Penrose, R. (1989). *The Emperor's New Mind*. Oxford: Oxford Univ. Press.

Pert, C. (1997). *Molecules of Emotion*. New York: Scribner.

Reber, A. (2018). *The First Minds*. Oxford: Oxford University Press.

Searle, J. R. (1992). *The Rediscovery of the Mind*. Cambridge, MA: MIT Press.

Shapiro, R. (1987). *Origins: A Skeptic's Guide to the Creation of Life on Earth*. New York: Summit Books.

Sheldrake, R. (1981). *A New Science of Life*. Los Angeles: Tarcher.

———. (1999). *Dogs That Know When Their Owners are Coming Home*. New York: Three Rivers Press.

Standish, L. J., L. Kozak, L. C. Johnson, and T. Richards. (2004). Electroencephalographic evidence of correlated event-related signals between the brains of spatially and sensory isolated human subjects. *The Journal of Alternative and Complementary Medicine* vol. 10, pp. 307–14.

Stapp, H. P. (1993). *Mind, Matter, and Quantum Mechanics*. New York: Springer.

Stevenson, I. (1987). *Children Who Remember Previous Lives: A Question of Reincarnation*. Charlottesville: Univ. Press of Virginia.

Taylor, G. R. (1983). *The Great Evolution Mystery*. New York: Harper & Row.

Teilhard de Chardin, P. (1961). *The Phenomenon of Man*. New York: Harper & Row.

Von Neumann, J. (1955). *The Conceptual Foundations of Quantum Mechanics*. Princeton: Princeton Univ. Press.

Wackermann, J., C. Seiter, H. Keibel, and H. Walach. (2003). Correlation brain electrical activities of two spatially separated human subjects. *Neuroscience Letters* vol. 336, pp. 60–64.

Waddington, C. (1957). *The Strategy of the Genes*. London: Allen & Unwin.

Wilber, K. (1981). *Up from Eden*. Garden City, NY: Anchor/Doubleday.

Wilber, K. (1993). *The Spectrum of Consciousness*. Wheaton, IL: Quest Books.

Wrangham, R. (2019). *The Goodness Paradox*. London: Profile Books.

INDEX

C

D

Professor, researcher, bestselling author, Quantum Science Pioneer, spiritual practitioner.

AMIT GOSWAMI, PHD, is a retired professor from the physics department of the University of Oregon (1968 to 1997). He is a renowned pioneer of the new paradigm of quantum science based on the primacy of consciousness.

In 2009, Amit started a movement called Quantum Activism, now gaining ground in North and South America (where he has founded Quantum Academy), Europe, and India. At the same time, he has established the Center for Quantum Activism (CQA) with headquarters in USA.

In 2019, he and his collaborators established an educational wing of CQA called Quantum Activism Vishwalayam (Home of the World), acting as Department of Quantum Science at the University of Technology in Jaipur, India, and developed a Master and PhD program in Quantum Science of Health, Prosperity and Happiness, an international program of transformative education.

Goswami has written several groundbreaking popular books based on research on quantum science and consciousness, amongst them, Quantum Integrative Medicine, Hero's Journey Quantum Style, The Quantum Brain, The Return of the Archetypes, Quantum

Spirituality, and The Re-enchantment of the Reality You Live (all of these, with Valentina R. Onisor, MD), Quantum Psychology and the Science of Happiness (with Sunita Pattani, MS), The Self-Aware Universe, The Quantum Doctor, Physics of the Soul, Quantum Creativity and The Everything Answer Book.

Amit was featured in the movie *What the bleep do we know?* and the documentaries *The Dalai Lama Renaissance* and *The Quantum Activist.*

He is a spiritual practitioner and calls himself a quantum activist in search of Wholeness.